D0206119

IUFRO Research Series

The International Union of Forestry Research Organizations (IUFRO), with its 14,000 scientists from 700 member institutions in 100 countries, is organized into nearly 300 research units that annually hold approximately 60 conferences, workshops and other meetings. The individual papers, proceedings and other material arising from these units and meetings are often published but in a wide array of different journals and other publications. The object of the IUFRO Research Series is to offer a single, uniform outlet for high quality publications arising from major IUFRO meetings and other products of IUFRO's research units.

The editing, publishing and dissemination experience of CABI *Publishing* and the huge spread of scientific endeavours of IUFRO combine here to make information widely available that is of value to policy makers, resource managers, peer scientists and educators. The Executive Board of IUFRO forms the Editorial Advisory Board for the series and provides the monitoring and uniformity that such a high quality series requires in addition to the editorial work of the conference organizers.

While adding a new body of information to the plethora currently dealing with forestry and related resources, this series seeks to provide a single, uniform forum and style that all forest scientists will turn to first as an outlet for their conference material and other products, and that the users of information will also see as a reliable and reputable source.

Although the official languages of IUFRO include English, French, German and Spanish, the majority of modern scientific papers are published in English. In this series, all books will be published in English as the main language, allowing papers occasionally to be in other languages. Guidelines for submitting and publishing material in this series are available from the Publisher, Books and Reference Works, CABI Publishing, CAB International, Wallingford, Oxon OX10 8DE, UK, and the IUFRO Secretariat, c/o Federal Forest Research Centre, Seckendorff-Gudent-Weg 8, A-1131, Vienna, Austria.

Forests and Landscapes
Linking Ecology, Sustainability and Aesthetics

Edited by

S.R.J. Sheppard

Departments of Forest Resources Management and Landscape Architecture, University of British Columbia, Vancouver, British Columbia, Canada

and

H.W. Harshaw

Department of Forest Resources Management, University of British Columbia, Vancouver, British Columbia, Canada

CABI *Publishing*

in association with
The International Union of Forestry Research Organizations
(IUFRO)

CABI *Publishing* is a division of CAB *International*

CABI Publishing
CAB International
Wallingford
Oxon OX10 8DE
UK

CABI Publishing
10E 40th Street
Suite 3203
New York, NY 10016
USA

Tel: +44 (0)1491 832111
Fax: +44 (0)1491 833508
Email: cabi@cabi.org
Web site: http://www.cabi.org

Tel: +1 212 481 7018
Fax: +1 212 686 7993
Email: cabi-nao@cabi.org

A catalogue record for this book is available from the British Library, London, UK.

Library of Congress Cataloging-in-Publication Data
Forests and landscapes : linking ecology, sustainability and aesthetics / edited by S.
Sheppard and H. Harshaw.
        p.   cm. -- (IUFRO research series ; 6)
    Includes bibliographical references.
    ISBN 0-85199-500-4 (alk. paper)
        1. Sustainable forestry. 2. Forest ecology. 3. Forest landscape management.
    I. Sheppard, Stephen, 1951– II. Harshaw, H. (Howard) III. Series.

SD387.S87 F65 2000
333.75'152--dc21   00-064183

Published in Association with:

The International Union of Forestry Research Organizations (IUFRO)
c/o Federal Forest Research Centre
Seckendorff-Gudent-Weg 8
A-1131 Vienna
Austria

ISBN 0 85199 500 4

Printed and bound in the UK by Cromwell Press, Trowbridge, from copy supplied
by the authors.

Contents

Image Credits

Photographs

Contributors

Simon Bell
Senior Research Fellow, School of Landscape Architecture, Edinburgh College of Art/Heriot Watt University
Edinburgh, UK

Simon Bell is currently a Senior Research Fellow at the School of Landscape Architecture in the Edinburgh College of Art/Heriot Watt University. Mr. Bell has had an active and varied career as the Chief Landscape Architect for the British Forestry Commission, a private consultant for government and industry, and author. Mr. Bell's recent book, *Landscape: Pattern, Perception, and Process* was published last year.

Daniel B. Botkin
Research Professor, Department of Ecology, Evolution and Marine Biology, University of California
Santa Barbara, USA

Daniel Bodkin is Professor of Biology, George Mason University, and President of the Center for the Study of the Environment, Santa Barbara, California. Dr. Botkin is well known for his ideas about non-steady-state characteristics of ecological systems, as expressed in his books, *Discordant Harmonies: A New Ecology for the 21st Century*, and *Our Natural History: The Lessons of Lewis and Clark.* Dr. Botkin has directed ecosystem research in many parts of the world, from the Serengeti Plains of Africa to forests of Siberia; his published research includes studies of forests, whooping cranes, salmon, bowhead whales, moose and African elephants. His work has emphasized a systems approach, including the study of life from a planetary perspective. In recent years, Dr. Botkin has focused on the application of environmental sciences to solve complex environmental problems.

Jeffery Burley
Director, Oxford Forestry Institute and President, International Union of Forestry Research Organizations (IUFRO)
Oxford, UK

Professor Jeff Burley has been the Director of the Oxford Forestry Institute in the University of Oxford for the past fifteen years; and is currently the President of IUFRO. Following an undergraduate education in forestry at Oxford, he obtained Masters and Doctoral degrees in forest genetics from Yale University. Dr. Burley has worked for five years as Head of the Forest Genetics Research Centre in central Africa, and was involved in projects in Zambia, Malawi, and Zimbabwe; he has worked as a consultant on projects in 35 countries. Dr. Burley has held three separate posts at Oxford, all of which have involved education, training, research and information for tropical countries and for international development institutions. As President of IUFRO, he has been responsible for the establishment of a number of Task Forces including one on Sustainable Forest Management that is seeking to marry the power of research and the validation of criteria and indicators for the sustainable management of forests.

Allen Carlson
Professor of Philosophy, University of Alberta
Edmonton, Alberta, Canada

Allen Carlson is a Professor of Philosophy at the University of Alberta. His teaching and research interests include aesthetics, environmental philosophy, with particular focus on environmental and landscape aesthetics. He has co-edited two collections of essays in this latter area and has published articles on the aesthetics of nature, on environmental appreciation, and on landscape assessment and evaluation. Dr. Carlson's research has appeared in various anthologies and in journals such as *The British Journal of Aesthetics*, *The Canadian Journal of Philosophy*, *Environmental Ethics*, *The Journal of Aesthetics and Art Criticism*, *Landscape Journal*, and *Landscape Planning*. He has a book on environmental aesthetics shortly forthcoming from Routledge Ltd.

John Danahy
Associate Professor, Centre for Landscape Research and the School of Architecture and Landscape Architecture, University of Toronto
Toronto, Ontario, Canada

John Danahy is an Associate Professor of Landscape Architecture with a cross appointment in Computer Science at the University of Toronto. He is the Director of the *Centre for Landscape Research* (CLR) and Director of Academic Computing for the Faculty of Architecture, Landscape and Design. The research laboratory that Danahy leads at CLR has played a leading role in the development of real-time visualization, design and decision support software since the mid 1980s. His work includes urban design for cities and simulation of large-scale landscapes for planning. CLR's research test-bed technology has been used in research and teaching activities by over 30 universities. An objective in Danahy's work has been to make visualization technologies accessible to more people, so that laypersons can effectively engage design experts in dialogue about the outcomes of spatial planning.

Terry Daniel
Professor, Environmental Perception Laboratory and Department of Psychology
University of Arizona
Tucson, Arizona, USA

Dr. Terry Daniel is a Professor of Psychology and Renewable Natural Resources at the University of Arizona. He received his M.A. in Psychology in 1968 followed by his Ph.D. in Experimental Psychology in 1969, both from the University of New Mexico. His research interests include environmental psychology, psychometrics and statistics, all of which focus upon parks, wilderness and other natural areas. Terry's work on the *Scenic Beauty Estimation Method* represents a major step forward in the quantification of non-market aesthetic values, and has been used to incorporate those values into the land management process. Along with his students, Terry applies classical psychophysical methods, psychometric models, computer visualization and survey techniques to determine how natural and human-designed changes in these environments affect the quality of experience for tourists, recreationists and residents. This work, and the methods that Terry has helped to develop are widely used to address the "human dimensions" of environmental protection and natural resources management. This work also provides theoretical insights into more general issues of human environmental perception and preference.

Howard W. Harshaw
Member of CALP, Department of Forest Resources Management, University of British Columbia
Vancouver, British Columbia, Canada

Mr. Harshaw's background includes studies in geography, outdoor recreation, parks and tourism. Currently, he is a graduate student at UBC and is examining the intrinsic and inherent values of forested landscapes to people and communities, with a particular focus on outdoor recreation and aesthetics. This work builds on Howard's past research, which has included an exploration of the motivational factors of canoe expedition participation, an examination of the role that public participation plays in natural resource and wilderness decision-making, and the development of a land-use inventory for the Rossport Islands in Lake Superior. Howard is a member of the *Collaborative for Advanced Landscape Planning* (CALP) at UBC, and plays an active role in undergraduate forestry recreation courses as a teaching assistant.

J.P. (Hamish) Kimmins
Professor, Department of Forest Sciences, University of British Columbia
Vancouver, British Columbia, Canada

Dr. Hamish Kimmins is an internationally known and respected forest ecologist. He is Professor of Forest Ecology in the Faculty of Forestry at UBC, and in the early 1990s served for two years as the Associate Director of the Sustainable Development Research Institute at UBC. He received his B.Sc. Degree from the University of Wales, Bangor, his M.Sc. from University of California, and a M.Phil. and Ph.D. in Forest Ecology from Yale. Over his 31 years of association with UBC he has been awarded numerous major awards, including the IUFRO gold medal for achievement in forest service (1985), the Canadian Institute of Forestry award for scientific achievement (1986), the Western Forestry and Conservation Award for Achievement in Forestry (1991) and the Industrial Forestry Lecturer Award, University of Alberta (1991). He frequently acts as consultant to federal and provincial government agencies, as well as international governments and organizations. Dr. Kimmins has published over 103 refereed journal articles and conference proceedings papers, as well as three books and nine book chapters. His textbook, *Forest Ecology* has become a standard text in many university's Forestry courses. *Balancing Act: Environmental Issues in Forestry* is used internationally in teaching and has become a UBC Press best seller. Both books are now in second editions. Hamish also remains a prominent figure in the public arena, frequently giving lectures within the community and participating in public debates on conservation issues.

Linda Kruger
Research Social Scientist, USDA Forest Service, Pacific Northwest Research Station
Seattle, Washington, USA

Dr. Linda Kruger works with the *People and Natural Resources Program* of the USDA Forest Service; her research focuses on community and forest relationships, and she is currently working on a problem analysis that will include a synthesis and a collection of papers on topics dealing with community-forest issues. Linda completed her Ph.D. at the University of Washington, College of Forest Resources in 1996; her dissertation examined the meanings of conceiving of place as a cultural system. Linda's interdisciplinary background has allowed her to examine and explore how the public, managers and scientists conceive of their environment and their landscapes, how people fit into these conceptions and what this interaction means for the sustainable management of resources and the role that the public can play in decision-making.

Don Luymes
Assistant Professor, Landscape Architecture Programme and Department of Forest Resources Management, University of British Columbia
Vancouver, British Columbia, Canada

Don Luymes' training in landscape architecture has given him broad experience in urban and environmental design, including projects in Toronto, Canada that sought to enrich the design of streets and sidewalks. Don was the principal author of the award winning *Healing an Urban Watershed* project on the restoration of the Don River in Toronto, and won a national design award for the Don Valley Brickworks master plan. While a faculty member at Pennsylvania State University he conducted a funded project for the City of Philadelphia on ways to improve the environments around community-based health-care facilities in North Philadelphia, and designed a new town near Beirut, Lebanon. Since arriving as a faculty member at UBC in 1996, Don has been involved in teaching design studio courses that focus on parks and on site planning. He has also been active in teaching Continuing Education courses at UBC and the BC Justice Institute, and in giving a number of public lectures on garden and urban design. Mr. Luymes' research interests focus on the interface between urban and natural or rural landscapes, and on issues of design representation. He has published papers on a wide range of topics in landscape architecture and environmental design, including gated communities, urban greenways, sustainable community design, design theory and critiques of a number of built landscapes

Chadwick D. Oliver
Professor, College of Forest Resources, University of Washington
Seattle, Washington, USA

Dr. Oliver is a Professor of Silviculture and Forest Ecology in the Management and Engineering Division of the College of Forest Resources at the University of Washington in Seattle, where he also served as a contributing faculty member of the *Centre for International Trade in Forest Products* (CINTRAFOR). Chadwick's research work is concerned with the understanding of the ecological aspects of how forests develop and the way in which silviculture can be applied to ecological systems most effectively. Dr. Oliver a co-author of *Forest Stand Dynamics* and was an editor of *Forest Pruning and Wood Quality of Western North American Conifers.*

Brian Orland
Professor, Department of Landscape Architecture, Pennsylvania State University University Park, Pennsylvania, USA

Brian Orland was founder, in 1985, of the *Imaging Systems Laboratory* at the University of Illinois and is currently a professor of Landscape Architecture at the Pennsylvania State University. His research interests include (i) the investigation of patterns of responses which relate peoples' subjective judgements of environmental quality to the physical, and manageable, characteristics of the environment; (ii) development of computer-based visualization systems as part of standard data analysis systems that can represent a range of environmental conditions sufficiently well to support decision-making as well as visual perception research; (iii) methodological studies is data visualization and (iv) development of computer software tools to support his research interests.

Stephen R.J. Sheppard
Associate Professor, Department of Forest Resources Management and the Landscape Architecture Programme
Director of the Collaborative for Advanced Landscape Planning
University of British Columbia
Vancouver, British Columbia, Canada

Dr. Sheppard holds a B.A. and an M.A. in Agricultural and Forest Sciences from Oxford, an M.Sc. in Forestry from UBC, and a Ph.D. in Environmental Planning from University of California at Berkeley. As a faculty member in both Forest Resources Management and Landscape Architecture, he is involved in social forestry and resource planning, and directs the Collaborative for Advanced Landscape Planning (CALP), specializing in research on computer visualization, user-friendly Geographic Information Systems and public response testing. He is an internationally recognized expert in visual resource management and visualization, and published in 1989 a key text entitled *Visual Simulation: a User's Guide for Architects, Engineers, and Planners.* Stephen has worked for over twenty years as a landscape architect and senior planner concerned with environmental and aesthetic issues of the landscape. Stephen's experience in the UK, USA, Canada and other countries has focused on aesthetics, public participation, and project decision-making.

David B. Tindall
Assistant Professor, Departments of Forest Resources Management and Sociology,
University of British Columbia
Vancouver, British Columbia, Canada

Dr. Tindall specializes in environmental sociology; his current research involves working on a set of interrelated research projects which examine values, attitudes, and behaviour related to the ongoing controversy over forestry and conservation in the old growth rainforests of British Columbia. These studies focus on people in environmental organizations, forestry organizations, and the general public in resource communities and urban centres. Aspects of this research include survey research on community members' values and concerns related to forest characteristics and forest management, and analyses of media coverage of forestry and conservation in British Columbia. Dr. Tindall is involved the Sociology, and the Natural Resource Conservation programmes at UBC, and teaches courses on Society and Natural Resources Perspectives on First Nations and Forest Lands.

Umeek (Dr. E. Richard Atleo)
Instructor, First Nations Studies, Malaspina University-College
Nanaimo, British Columbia, Canada

Dr. Atleo earned his doctorate from UBC and is currently developing and instructing the First Nations Studies undergraduate programmes at Malaspina University-College; he is also involved in the development of a graduate programme in First Nations Studies. Dr. Atleo has published in the areas of education, anthropology, museology, environmental ecology, and political ecology. His research involvement has ranged from education to medical science to ethno-ecology. From 1993 to 1995, Dr. Atleo served as Co-Chair of the Scientific Panel for Sustainable Forest Practices in Clayoquot Sound. The recommendations of this panel for the implementation of sustainable forestry practices in Clayoquot Sound were accepted by the British Columbia Provincial Government.

Jori Uusitalo
Senior Lecturer of Wood Science, Faculty of Forestry, University of Joensuu
Joensuu, Finland

Dr. Jori Uusitalo is senior lecturer of Wood Science at the University of Joensuu. After his dissertation at the University of Helsinki in 1995 he conducted post-doctoral study in forest visualization at the *Imaging Systems Laboratory*, University of Illinois under the supervision of Professor Brian Orland. He currently coordinates several research projects that deal with lumber quality prediction, pre-harvest measurement, forest visualization and computer-aided bucking of harvesters.

JoAnna Ruth Wherrett
Member, Land Use Science Group, Macaulay Land Use Research Institute
Craigiebuckler
Aberdeen, UK

JoAnna completed her Ph.D. studies in landscape research in 1998, undertaken jointly at the Macaulay Land Use Research Institute and the Robert Gordon University in Aberdeen. Prior to this she gained degrees in Mathematics from Cambridge University, England and in Rural Resource Management from the University College of North Wales in Bangor, Wales and worked for the Scottish Office Agriculture, Environment and Fisheries Department and the British Trust for Conservation Volunteers. JoAnna has worked as a research assistant at the Macaulay Institute, and has spent time as a visiting Post-doctoral Researcher at the University of Joensuu, Finland, studying landscape preference in the Finnish Lake Region. She is currently working as a Research Fellow in Landscape Perception at the Edinburgh College of Art/Heriot-Watt University, in Edinburgh, Scotland. Research interests include the use of electronic media for gathering information on landscape preference and perception, the influence of cultural and historical factors in landscape perception and the association between landscape sustainability, biodiversity and scenic quality.

Foreword

Paul H. Gobster
North Central Research Station, USDA Forest Service
Chicago, Illinois

The prairies of the Midwestern US seem like an unlikely place for me to comment on the fruits of a workshop inspired by the forests of Canada's Pacific Northwest. But it is from this same vantage point, more than a half-century earlier, that Aldo Leopold outlined the elements of what has since been called an *ecological aesthetic*, a way of beholding the landscape that the contributors to this book consider as one of several means for reconciling apparent conflicts between our preferences for landscape as scenery and our desire for those landscapes to be ecologically sustainable. In examining the philosophical, scientific, and pragmatic dimensions of linking aesthetic and sustainability values, participants in the Peter Wall Exploratory Workshop have gone far to advance our understanding of Leopold's ecological aesthetic, both exciting us about its possibilities and cautioning us about its limitations. As you delve into the chapters of this book, I am sure that you will receive not only many good answers on how to deal with aesthetic and sustainability values, but will also be instilled as I was with the urge to question, test, refine, and apply the ideas and suggestions the contributors have put forth.

My task to place the context and significance of this effort is an easy one, for I believe this book proclaims no less than a second revolution in the way we as a society think about, study, and act upon the aesthetic dimensions of forest landscape management. The first revolution, driven by controversies over clearcutting in the 1960s, led to explicit recognition of aesthetic values in laws and practices governing the management of forest landscapes, notably the *National Forest Management Act* in the US and the USDA Forest Service's *Visual Management System*. The Romantic view of nature-as-scenery, with its emphasis on the dramatic, visual, and static elements of landscape, served as our first aesthetic model for landscape planning and research. Formal qualities such as unity, vividness, and variety in line, colour, form, and texture, taken from aesthetic theories in the visual arts as interpreted through the Romantic schools of landscape painting and naturalistic landscape design, served as evaluation criteria in expert-based landscape assessments. These assessments helped to protect the most scenic areas from timber harvesting and instructed forest managers on how to mitigate visual impacts elsewhere by leaving vegetation screens along roadsides and undulating the edges of clearcuts. While studies of public preference added further depth and understanding to the expert assessments, most research was also conducted within the paradigm of the scenic model, for instance asking individuals to make rapid perceptual ratings of the scenic quality of landscapes framed and presented as photographs taken at one point in time. By eliminating the extra-visual and temporal dimensions of landscapes and by focusing on only the immediate perceptual component of people's aesthetic responses, researchers helped confirm scenic ideals of landscapes as showy and undisturbed by natural processes or human interventions.

Manifestations of a second revolution in understanding and dealing with the aesthetics of forest landscapes came in North America in the late 1980s and early 1990s with the spotted owl controversy in the Pacific Northwest, increased public recognition of biodiversity as a forest resource, and adoption of ecosystem management policies by public land management agencies. Forests would need to be managed differently to provide these sustainability values, and in many cases prescriptions to improve such things as ecosystem health and biodiversity would result in a much different look to the forest.

I first became aware of the aesthetic implications of such a shift at a regional meeting of US Forest Service landscape architects in Milwaukee, Wisconsin in the autumn of 1990. At that meeting we heard a talk by the Regional Silviculturist about New Perspectives, an ecosystem-based approach to management that had recently been adopted by the Forest Service. The silviculturist remarked that we as landscape architects would be natural leaders in implementing the programme because of our sensitivity to public "white hat" values like aesthetics and biodiversity. He then went on to describe a New Perspectives project proposed for northern Wisconsin, where they planned to create a series of 900-acre openings in the national forest to provide habitat for moose and elk reintroduction. In a region where landscape architects fought to keep clearcuts under 20 acres, this large block management was unheard of and would directly conflict with visual quality objectives specified under the Visual Management System.

In the following months I came across additional examples of *when worlds collide*, conflicts between practices developed to maintain scenic quality and those that would protect or enhance sustainability values. These clashes related not only to timber harvesting under New Perspectives, but also to ecological management in ecosystem restoration where no obvious utilitarian goods were being extracted from the forest. How could such conflicts be resolved given that both aesthetics and ecological sustainability were seen as noble, *white hat* goals? Perhaps the problem was not so much that biodiversity and ecosystem health were incompatible with aesthetics, but that aesthetics as it was being conceived, measured, and addressed in landscape planning was not given a fuller reading.

Aesthetic philosophy has in some ways been responsible for the problems inherent in the scenic model of landscape management, yet a new wave of environmentally oriented aesthetic philosophy now coming of age could provide fresh guidance in dealing with perceived conflicts between aesthetic and sustainability values. Among the earliest progenitors of this wave, Ronald Hepburn (1968) described important ways in which the aesthetics of nature and landscape differed from that of art and the built environment, and Allen Carlson (1977) argued that methods to plan for and manage natural landscapes suffered from the biases inherent in an inappropriate scenery model.

It was Baird Callicott (1983) who first began to piece together Leopold's "ecological and evolutionary land aesthetic" from essays in *A Sand County Almanac* (Leopold, 1949), and he and Susan Flader (1991) subsequently compiled Leopold's ideas on ecological aesthetics from his various writings through the years. As an alternative to the dominant scenic aesthetic, Leopold's ecological aesthetic expands our goal of identifying and protecting the *most scenic* landscapes to one aimed at discovering the beauty that lies *within each* landscape. Rejecting the Romantics' formal notions of beauty, Leopold looked to the ideas of ecological integrity and health as guides to aesthetic appreciation. Such an appreciation relies as much on our understanding about science and the workings of nature as on our visceral reaction to the sights, sounds, and smells we experience.

Philosophy related to how we think about aesthetics in the context of nature, ecosystem health, and the management of forest landscapes has burgeoned in recent years, with important writings by Carlson (*e.g.* 1993; also see Chapter 3 in this volume), Marcia Eaton (1997), Cheryl Foster (1998), Holmes Rolston III (1998), Yuriko Saito (1998), and others. Those working in ecological design, planning, and research in non-forest contexts also have much to offer, including Catherine Howett (1987), Anne Whiston Spirn (1988), Robert Thayer (1989), Joan Nassauer (1997), and others. These ideas are bringing about a second revolution in our thoughts and practices dealing with the aesthetics of landscapes, one that offers potential in resolving apparent conflicts between aesthetic and sustainability values.

In 1992 when I first wrote about applying these ideas in a forest management context (Gobster, 1996), I concluded that although I thought an ecological aesthetic was a good idea whose time had come, it would be difficult to adopt directly because the scenic aesthetic was so firmly entrenched in our cultures of research, practice, and society at large. Instead of trying to change this deeply held value, I suggested that as a short term strategy managers and researchers should take a more indirect approach, asking people whether they found particular ecosystem management goals and practices socially acceptable or appropriate when given knowledge about the context or setting in which such management would take place. In subsequent papers (Gobster, 1994; 1995; 1997; 1999) I tried to suggest how one might actually help realize an ecological aesthetic over the long term as it applies to planning and policy development, on-the-ground management, and research and theory development. Along the way, I have received both support and criticism that have helped refine my thinking and which, in ways unknown to me, may have in part led to this book.

The editors and contributors to this volume have gone far beyond my initial attempts at synthesis and application, and their efforts are evidence that the second revolution is well on its way. If there is anything that I can add to their efforts from my own experience it is a note of cautious encouragement based on some of the areas where I myself have stumbled. To conclude, I offer the following observations on how we might better proceed in linking aesthetics and sustainability values.

Avoid Either/Or Constructions of the Issues

In past papers I have highlighted some key areas where a scenic model of landscape aesthetics might conflict with sustainability values, and have then gone on to identify some ways in which an ecologically-based model might sidestep or help resolve these problems. Some people have said in doing so I have overplayed the conflict between what is scenically beautiful and what is ecologically sustainable, and have set up the scenic model as a straw man to be knocked down by a *superior* ecological one. While I continue to think that there can be real problems in taking a strict scenery approach to management, I also recognize one can go too far the other way as well. Certainly there are many cases where scenic beauty and biodiversity go hand in hand, and likewise there are many attributes of scenic landscapes (and our responses to and interactions with them) that we would be worse off without. By building upon our ideas of what is beautiful and by incorporating ideas of ecological beauty, we can expand rather than substitute the forest landscapes, conditions, and processes we love, care about, and, ultimately, are willing to protect. I think this is the middle ground that Joan Nassauer has identified in her research on *cues to care* (*e.g.* Nassauer, 1995), and what Stephen Sheppard (Chapter 11, this volume) arrives at in his theory of visible stewardship. More information on this middle ground is needed, not only from theorists and researchers, but from practitioners as well. As one US Forest Service landscape architect told me in an impassioned letter about how things are worked out in the field in contrast to how I had portrayed them based on planning manuals and research papers:

> I have handled each of these "conflicts" in my work and have found them all to be very superficial, unnecessary and generally evaporate when LA and biologist work together to achieve a beautiful and functional desired future condition.... We don't use screens or hide ugly things anymore and we are not ashamed of some charcoal in areas that require underburning. Creation of edge is a small matter if you have already made certain to protect the greater share of interior habitat.... I feel the conflicts you mention are already outdated, old hat stuff. We look for the most biologically diverse landscape to shoot for and keep it easy on the eyes,

too (Diana Ross, personal communication, Bear Spring RD, Mt. Hood NF, Maupin, OR, December 29, 1992).

Think About Aesthetics and Sustainability Within a Broader, Multi-Value Framework

The values people hold for forest landscapes are diverse, and may not always be compatible. Just as our idea of aesthetics can and should be expanded, so too should we expand our idea of sustainability as it applies to forest landscape management. In a recent controversy over ecological restoration of prairie and savanna ecosystems in the Chicago area where I live, restoration proponents argued that it was necessary to remove exotic trees and conduct periodic controlled burns to return the landscape back to its former ecological integrity and native biodiversity (Gobster, 2000). Opponents countered that they not only preferred the existing, more densely forested conditions for aesthetic, recreation, and privacy/ solitude values, but argued that such conditions also maintained air quality and moderated urban heat island effects better than the more open, fire-dependent native landscape. Instead of arguing whose values are better, perhaps a more constructive way to proceed is to respect the legitimacy of these multiple values and work together to integrate them to achieve the shared goal of a sustainable future for nature and people. This is not an easy thing to accomplish, but is increasingly necessary in a multi-value, multicultural society. In this same vein, several of the authors in this book seek to expand ideas of sustainability to go beyond ecological considerations; in particular Linda Kruger's place-focused collaborative approach to identifying landscape meanings and negotiating multiple and conflicting values seems promising.

Move Beyond Preference Approaches to Aesthetic Assessment

As a social science researcher, I was trained in various *preference* approaches to understanding landscape aesthetics, to conceptualize aesthetic response as an immediate perception to landscape stimuli. In contrast to this view, most contemporary philosophers of landscape aesthetics focus on the idea of *appreciation*, in which knowledge, experience, and learning play important roles. Because the cognitive dimension is such an important part of an ecological aesthetic, I think researchers and practitioners would do well to pay more attention to ideas inherent in the aesthetic appreciation of landscapes. Again, we should not aim to replace one idea with the other, but greater emphasis put on understanding, measuring, and providing opportunities for people to learn about and appreciate sustainable ecosystems could lead to expanded ideas of landscape beauty. In this respect, some of the *visualization* techniques described in this volume show promise for adapting to questions about appreciation. In other cases, we will need to expand our repertoire of approaches to better capture the multi-sensory, cognitive, and experiential qualities of landscape.

Is There a Place for Normative Theory in Landscape Research and Practice?

A final question I would like to raise to this book's readers and contributors concerns the role of researchers and practitioners in establishing a normative or prescriptive content in landscape aesthetic assessment. An ecological aesthetic clearly incorporates an ethical dimension; it implies that landscapes that are managed to increase health and diversity should be appreciated over those that compromise these sustainability values. This is different from a typical preference approach that attempts only to describe what people like, ostensibly making no judgment whether these preferences have any impact good or bad on the landscape. Should we, as social scientists and practitioners, do all we can to maintain our neutrality and just stick to the facts, or should we join our colleagues in the arts and humanities and advocate movement toward a more ethically based foundation for our work?

I see no reason why we cannot make such a move and maintain integrity in our science and practice. As any postmodern observer would instantly recognize, we impose our value structure on a problem the minute we begin to address it: for example, whether we choose to study preferences or appreciation, how we define and measure ambiguous concepts such as beauty, and how we "scientize" value-laden concepts like sustainability, ecosystem heath, and biodiversity (Hull and Robertson, 2000). In fact, the more we make clear our values and biases, the better position we are in providing answers to questions that withstand the rigours of public and professional critique. We have much to learn from those aesthetic philosophers and ecological designers who are working on related problems and issues, and it would behoove us to develop closer collaborations with them to infuse our research and practice.

These, then, are some of the opportunities and challenges that lie ahead as we pursue answers to vexing questions about aligning aesthetic and sustainability values. It is with their improvisations upon Aldo Leopold's normative ethic of landscape appreciation, and their explorations beyond it, that the contributors to this volume are leading us toward the next revolution in building landscape aesthetics into sustainable forest management.

References

Callicott, J.B. (1983) Leopold's land aesthetic. *Journal of Soil and Water Conservation* 38: 329-332.
Callicott, J.B., and S. Flader (eds) (1991) *The River of the Mother of God and*

Other Essays by Aldo Leopold. University of Wisconsin Press, Madison.

Carlson, A. (1977) On the possibility of quantifying scenic beauty. *Landscape Planning* 4: 131-171.

Carlson, A. (1993) On the theoretical vacuum in landscape assessment. *Landscape Journal* 12(1): 51-56.

Eaton, M.M. (1997) The beauty that requires health. In: J.I. Nassauer (ed) *Placing Nature: Culture and Landscape Ecology*, pp. 85-106. Island Press, Washington, DC.

Foster, C. (1998) The narrative and the ambient in environmental aesthetics. *The Journal of Aesthetics and Art Criticism* 56(2): 127-137.

Gobster, P.H. (1994) The aesthetic experience of sustainable forest ecosystems. In: W.W. Covington and L.F. DeBano (tech. coords.), *Sustainable Ecological Systems: Implementing an Ecological Approach to Land Management* July 13-15, 1993, Flagstaff, Arizona (Gen. Tech. Rep. RM-247), pp. 246-255. Ft. Collins, CO: USDA Forest Service, Rocky Mountain Forest and Range Experiment Station.

Gobster, P.H. (1995) Aldo Leopold's ecological esthetic: Integrating esthetic and biodiversity values. *Journal of Forestry* 93(2): 6-10.

Gobster, P.H. (1996) Forest aesthetics, biodiversity, and the perceived appropriateness of ecosystem management practices. In: M.W. Brunson, L.E. Kruger, C.B. Tyler, and S.A. Schroeder (eds), *Proceedings: Workshop on Defining Social Acceptability of Forests and Forestry Practices*, June 23-25, 1992, Kelso, WA (Gen. Tech. Rep. PNW-369), pp. 77-98. USDA Forest Service, Pacific Northwest Research Station, Seattle, WA.

Gobster, P.H. (1997) The Chicago wilderness and its critics III. The other side: A survey of the arguments. *Restoration and Management Notes* 15(1): 33-38.

Gobster, P.H. (1999) An ecological aesthetic for forest landscape management. *Landscape Journal* 18(1): 54-64.

Gobster, P.H. (2000) Restoring nature: Human actions, interactions, and reactions. In: P.H. Gobster and R.B. Hull (eds) *Restoring Nature: Perspectives from the Social Sciences and Humanities*, pp. 1-18. Island Press, Washington, DC.

Hepburn, R. (1968) Aesthetic appreciation of nature. In: H. Osborne (ed), *Aesthetics in the Modern World*, pp. 44-66. Thames and Hudson, London.

Howett, C. (1987) Systems, signs, and sensibilities: Sources for a new landscape aesthetic. *Landscape Journal* 6(1): 1-12.

Hull, R.B., and D.A. Robertson (2000) The language of nature matters: We need a more public ecology. In: P.H. Gobster and R.B. Hull (eds), *Restoring Nature: Perspectives from the Social Sciences and Humanities*, pp. 97-118. Island Press, Washington, DC.

Leopold, A. (1981) (originally published 1949) *A Sand County Almanac and Sketches Here and There*. Oxford University Press, New York.

Nassauer, J.I. (1995) Messy ecosystems, orderly frames. *Landscape Journal* 14(2): 161-170.

Nassauer, J.I. (1997) Cultural sustainability: Aligning aesthetics and ecology.

In: J.I. Nassauer (ed), *Placing Nature: Culture and Landscape Ecology*, pp. 65-84. Island Press, Washington, DC.

Rolston, H. (1998) Aesthetic experience in forests. *The Journal of Aesthetics and Art Criticism* 56(2): 157-166.

Saito, Y. (1998) The aesthetics of unscenic nature. *The Journal of Aesthetics and Art Criticism* 56(2): 102-111.

Spirn, A.W. (1988) The poetics of city and nature: Towards a new aesthetic for urban design. *Landscape Journal* 7(2): 108-126.

Thayer, R.L. (1989) The experience of sustainable landscapes. *Landscape Journal* 8(2): 101-110.

Acknowledgements

This book is the result of the efforts of many. The editors are grateful for the help and assistance provided by all involved with this endeavour.

The Peter Wall Institute for Advanced Studies and the Faculty of Forestry at the University of British Columbia provided the financial assistance for the organization of the workshop that initiated this volume, *Linking Forest Sustainability to Aesthetics - Do People Prefer Sustainable Landscapes?*, and that brought all of the participants together.

Hamish Kimmins of the Department of Forest Sciences and Kellogg Booth of the Media and Graphics Interdisciplinary Centre (MAGIC), both at the University of British Columbia, were instrumental in providing this volume with a colour graphics section.

Members of the Collaborative for Advanced Landscape Planning (CALP) were supportive throughout the development of this book, especially Jonathan Salter for his assistance with the colour graphics and his keen eye; Duncan Cavens for his design sense and technical knowledge; John Lewis, for providing some of his computer simulations; and Cecilia Achiam, for her help with the graphic layout, and her keen sense of design.

Finally, the editors would like to acknowledge the participants of the Peter Wall Workshop for engaging in lively and thought provoking discussions of the relationships between aesthetics and sustainability that provided the enthusiasm and impetus for this book.

PART I

Linking ecological sustainability
to aesthetics: Do people prefer
sustainable landscapes?

Chapter one:

Landscape Aesthetics and Sustainability: An Introduction

Stephen R.J. Sheppard and Howard W. Harshaw
Department of Forest Resources Management and the Centre for Advanced Landscape Planning, University of British Columbia
Vancouver, British Columbia

1 Part I

1.1 Issues and Questions

The public tends to judge forest management by how it looks. What happens then when forest ecologists and resource managers seek to implement ecosystem management strategies and new forest practices intended to make forestry more sustainable? Will the public like what they see? Will they understand it? Will they accept it? In short, do people prefer sustainable landscapes over less sustainable ones?

As the forestry profession and resource managers in general seek to bring their activities more into line with natural processes, under the influence of scientific, political, and public pressure on a local and global scale, these questions assume major importance. Certification schemes for forestry and land conservation seem to be the current great white hope of the resource industry in securing market share and achieving what has been termed a "social license" to harvest resources. Ecosystem management may sound fine to everyone as an overall strategy, but what if the implementation on the ground fails to measure up to popular expectation? What happens when the public becomes upset at the appearance of a certified sustainable forest?

Current conflicts in places like the Elaho Valley near Vancouver, British Columbia - the scene of blockades, violence, and much controversy over timber harvesting in old growth and second growth forests - will become test cases for these questions. It is, however, already clear that public perception tends to equate visual degradation of landscapes with unsound - and, by implication, unsustainable - management practices. Other relationships between forest ecology, landscape aesthetics, and sustainability are less clear.

Forest scientists and resource managers seem to be divided between those who see a strong association between ecological health and visual quality, and those who do not. Forest design literature abounds with claims that steps to improve ecological conditions also improve aesthetic qualities, and vice versa (as described by Sheppard in Chapter 11 of this volume). As a practical example, the British Columbia Ministry of Forests (BCMoF) believes that timber supply at a Provincial level is overly constrained by the simple addition of ecological constraints from the Forest Practices Code and Visual Quality Objectives (VQOs), since these different constraints on timber harvesting would be expected to overlap geographically: ecological leave-strips and wildlife tree patches should also help significantly to attain VQOs (BCMoF, 1997).

The opposite view is held by those, especially among the forest sciences fraternity, who see sustainability as too complex to be directly related to visual landscape indicators, or to be assessed by a visual analysis approach. How can long-term subterranean soil processes, foliage nutrient levels, carbon content, or population numbers of nocturnal or secretive forest creatures, for example, possibly be related to aesthetic quality?

One of this book's primary objectives is to make some sense out of these opposing views. In this volume, we begin to map out some of the patterns in the relationships between ecology, sustainability, and aesthetics: in short, *are scenic landscapes ecologically sustainable* (and if so, under what circumstances), and *are ecologically sustainable landscapes scenic* (and if so, under what circumstances)? Put even more succinctly, *if it looks good, is it good, and if it looks bad, is it bad?*

The authors in this volume present and examine a range of views on the underlying theories, research needs, and practical approaches for managing the combination of ecological and aesthetic values, as well as some of the difficulties of dealing with public knowledge, perceptions and preferences. For instance, should the public be better educated to understand that what looks bad may actually be good? Can a new aesthetic be developed to incorporate our current understandings of sustainability? What are the ethical concerns of seeking to impose a new aesthetic?

If we are successful in identifying the relationships between aesthetics and ecological sustainability, what are the implications? Why does all this matter? Without a clearer resolution of this dilemma, progress toward sustainable forestry may be seriously hampered by scientific and professional confusion over strategies for promoting sustainability to the public. Some scientists would argue that public perceptions rooted (at least partly) in past cultural norms are not reflective of current global priorities for ecological sustainability, and already stymie honest attempts to steward the forest appropriately. Is the practice of sustainable forestry itself sustainable, without public understanding and support? Certainly, those whose living, tax-base, or profitability depend on validation of sustainable forest practices will be adversely affected by public rejection of those practices.

On the other hand, given the arguments among the scientific and land management professions on the definition of sustainability, and the inadequacy of our knowledge on the long term health of the forest environment, it is possible that the so-called sustainable practices of today may turn out to be less desirable than we think. Will it turn out that the original public aesthetic was ecologically correct all along (at least in some situations?). What if practices that are generally assumed to be good stewardship and therefore sustainable, such as tree-planting and pest control, actually receive public approval mainly because they look good (to both foresters and the public), rather than on a sound scientific rationale? For example, the visible restoration of the forest by the replanting of clearcuts in the Pacific Northwest meets with general public acceptance - rapid "green-up" is good - despite the fact that uniform monocultures of non-local genotypes of Douglas Fir are commonly used to maximize long-term timber supply and future profits, rather than to achieve clear ecological goals.

Discussion of issues and ideas such as these was the raison d'être for an international workshop (described next), which formed the genesis of this volume.

1.2 Linking Forest Sustainability to Aesthetics: The Peter Wall Institute Exploratory Workshop

In the early spring of 1999, an interdisciplinary panel of leading ecologists, forest resource scientists, landscape architects, philosophers, perceptual psychologists, sociologists, and computer scientists gathered together at an Exploratory Workshop held at the University of British Columbia (UBC) in Vancouver, Canada. Sponsored by the Peter Wall Institute for Advanced Studies (PWIAS) and the Faculty of Forestry at UBC, with support from IUFRO and other organizations, the purpose of the Workshop was to debate and explore, for the first time at this level, relationships between ecology and public preferences, and to examine the interactions between aesthetic quality and sustainability of forest resource management. The Workshop also explored the potential of computer visualization of forested landscapes to cross disciplinary boundaries in research on perceptions of sustainability, and perhaps to help resolve some of the difficult issues identified in the debate.

The establishment of the Peter Wall Institute for Advanced Studies has permitted interdisciplinary research groups led by UBC members to explore and examine fundamental issues concerning the advancement of knowledge. The focus of the research efforts is that of basic research rather policy oriented. The exploratory workshops that are made possible by the PWIAS allows distinguished external experts and UBC researchers to jointly work toward assessing the research possibilities in a new area and to the development of a research agenda.

The four-day Exploratory Workshop stimulated a lively debate between disciplines which rarely sit down together to hammer out forest landscape issues. It also turned the spotlight on to some of the human dimensions issues of forest ecology and management which often receive less thorough academic and professional attention in forestry circles. Its intended outcomes included a discussion of research needs and approaches to help resolve some of the complex theoretical and practical questions.

The Workshop format was varied to facilitate open communication amongst the participants and to provide for an opportunity for discussion with the interested public. The Workshop was divided into moderated sessions that began with lectures from participants and were followed by open discussions. A public day allowed for people from academia, industry and private practice to interact with the Workshop participants and view demonstrations of visualization-based decision tools for forested landscapes (Peter Wall Exploratory Workshop, 1999).

1.3 Scope and Definitions

Before launching into in-depth consideration of the relationships between sustainability and aesthetics, it is important to define the scope of the book and some key terms used (fairly consistently) throughout. It is also hoped that it would be helpful to those readers who are new to some of these subjects to provide a general introduction to some of the aesthetic and ecological dimensions of forest management to be discussed in the following pages.

The discussion in this volume is focused on forest resources and forested landscapes, although many of the arguments are applicable in some way to other types of broader working landscapes and natural resource/land management activities. The scope of the discussion on forestry is international, and not restricted to the Pacific Northwest or Canada, though several chapters are illustrated with several examples from the front-line of forestry in these and other locations, wherever issues of aesthetics and sustainability are debated.

The book is intended to offer something for resource managers, scientists and lay people interested in a socially sensitive and environmentally responsible form of forestry. The intended readers include professionals grappling with real world problems, who may be able to distil new approaches and techniques from these chapters, and scientists in various disciplines interested in human dimension research and the interplay of ecological and social factors in forestry practice. Policy-makers may find fresh insights and new perspectives on the practical dilemmas stemming from differing perceptions of forest values. Moreover, the book would be of interest to anyone concerned with the aesthetics of sustainability, in any land use type or context.

The central notion of *sustainability* and what it can mean is defined in several chapters, notably those by Daniel, Oliver *et al.*, Burley and Kruger. For the purpose of this introduction, it is sufficient to note that the primary use of the term in this book focuses on ecological sustainability, rather than the broader definition of economic and social sustainability; nonetheless, the importance of these other aspects of sustainability is not ignored, and is addressed in most depth by Kruger in *Part III*.

The term *aesthetics* is taken in its broadest sense, to encompass much more than just visual quality: it is used here to express the full range of aesthetic and perceptual qualities received by the senses and appreciated by the mind, including the meanings to be found in the landscape, such as the symbology of timber harvesting practices and forest stewardship. More specific definitions and explanations are provided by several authors, particularly Daniel, Carlson and Bell. The term *perception* can be used in two ways: as sensory perception, *i.e.* the mechanisms of vision; or more loosely, as opinion or preference based on an understanding of the characteristics of a situation. *Preference* is again a more specific term (see also Daniel in Chapter 2) for the degree to which a person or group prefers an situation or feature over other situations or features; used here, it generally applies to aesthetic and related preferences for forest conditions and landscapes, not the full range of possible types of preferences which could be explored in human dimension research. While we can expect some correspondence between landscapes which are aesthetic and those which are preferred, there can be many other reasons for overall preference, beyond its aesthetic qualities.

The term *landscape* can have some distinct meanings, depending on the discipline of the reader. To a forester, it refers to a geographic scale above that of the forest stand, but below that of the region: it often coincides with medium or large watersheds, and incorporates all the elements of the ecosystem. In landscape architecture, the word *landscape* more explicitly includes consideration of the aesthetic (and often visual) characteristics of the environment, sometimes taken together with the biophysical conditions.

1.4 Structure of this Volume

This volume therefore presents a unique collection of papers by distinguished authors interested in exploring the relationships between forest ecology and aesthetics. New research on public attitudes to forestry and ecological models for sustainability are combined with reviews of sustainability indicators which address amenity values, and the clash between visual resource management and emerging theories of an ecological or stewardship-based aesthetic. New techniques of virtual forest visualization that convey the complexities of spatial and temporal landscape change are illustrated by some of the world's experts in this rapidly growing field.

The remainder of this book is structured in five parts, as described next.

2 Part II: Seeing and Knowing: Approaches to Aesthetics and Sustainability

Part II refines the questions posed by the relationships between aesthetic considerations and ecological sustainability of the forest, and sets out some basic arguments on, and understandings of, these potential relationships.

In Chapter 2, Daniel addresses the central topic of the book in terms of the ecological aspects of sustainability and the largely visual aspects of aesthetic preference. In seeking to determine the psychophysical basis for aesthetic preferences (*i.e.* human responses to landscape stimuli), psychologists typically differentiate two principle mechanisms: expressed preferences as the outcome of rational processes based in part on environmental information (*e.g.* information on sustainability); and expressed preferences resulting from reflexive emotional responses to environmental stimuli. The degree to which the one can overpower the other in responding to forest management activities is a crucial question. Daniel goes on to consider a number of fundamental and operational challenges to be overcome in attempting to resolve relationships between aesthetics and ecological sustainability, themes that are developed later in the book by several of the authors.

Carlson, in Chapter 3, further distinguishes the differences between preferences in general, aesthetic preferences, and aesthetic value by redefining the Exploratory Workshop research question as: "Do people prefer *the look* of sustainable landscapes?". His philosophical argument suggests that people may indeed aesthetically prefer sustainable landscapes, by explaining how knowledge of a landscape's nature (*e.g.* its ecological status) informs the perception of it: two similar landscapes that have different natures will strike us as aesthetically different.

Aldo Leopold (1968), referring to largely deforested, agricultural landscapes, noted that if something looks bad, then it probably is "bad" ecologically. Kimmins, in Chapter 4, suggests that this simple maxim has been widely embraced, but is in conflict with ecological realities in many forested landscapes. Although visual landscape quality is an important forest value, it must be balanced by the ecological requirements of the other values we wish to sustain. Kimmins argues that there are no simple relationships between readily visible indicators of forest ecosystem condition and various measures of sustainability.

Some of the connections between aesthetic values and other human and social values, and the relationship between these values and attitudes, are illustrated and explored by Tindall in Chapter 5. Drawing on recent original survey research, Tindall examines the human values associated with forests, and attempts to define who "the public" is. By understanding intergroup differences regarding aesthetic and other value preferences, it is suggested that we can begin to discern why certain forestry issues, such as hiding forestry behind leave strips and especially the practice of clearcutting, lead to such rancorous debate over what is appropriate for sustainable forest management.

3 Part III: Perspectives on Forest Sustainability

This section of the book delves deeper into our understanding of sustainable forestry, with fresh perspectives on current criteria and indicators used in determining levels of sustainability, and some suggestions for more robust, appropriate, and inclusive standards.

Values derived from forested landscapes are dynamic. Oliver *et al.*, in Chapter 6, question whether the evolution in perceptions of desired forest values, which has resulted in the reduced importance of traditional Western values (such as timber supply and employment), has significantly enhanced emerging forest values such as biodiversity. This chapter presents a theoretical framework for measures and perceptions of forest sustainability, and promotes a systems management approach that allows for the dynamic nature of forest values to be incorporated while balancing the range of desired values. Oliver *et al.* argue that the current debate over forest values is not so much concerned with what values to provide, as it is with how these values are to be provided.

A number of international attempts have been made to address the management of sustainable forests and to balance this with the provision and maintenance of human welfare. *Agenda 21*, the *Rio Declaration* and the *Forestry Principles*, developed at the *United Nations Conference on Environment and Development* in 1992, are examples of such attempts. In Chapter 7, Burley compares international forest management initiatives in order to indicate the requirements for identifying criteria and indicators of forest sustainability, and proposes that quantifiable indicators be included in the political processes. More specifically, Burley also examines the way in which these criteria and indicators address public preferences and aesthetics, and outlines possible approaches to incorporating amenity or landscape values into these sustainability frameworks.

As suggested in earlier parts of this volume, the influence of culture on aesthetic appreciation (both seeing and knowing) cannot be denied. However, this influence is generally discussed in terms of Western concepts of beauty. Umeek (Dr. Richard Atleo) presents another approach to perceptions and ecology in Chapter 8. The Tloo-qua-nah principle of the Nuu-chah-nulth Nation is to remember reality, that "everything is one". From a First Nations perspective then, the question is not whether aesthetics or sustainability is preferred, but whether destructive harvesting practices are effectively replaced by more creative and respectful forest practices. Atleo argues that within the unity of existence, physical and spiritual forces are in constant opposition, and must be managed, balanced, and harmonized in order to enhance life within both the ecosystem and the culture dependent upon it.

4 Part IV: Theories Relating Aesthetics and Forest Ecology

Part IV turns from the consideration of sustainability itself to an in-depth review of relevant aesthetic theories on preferences for forested landscapes, as they relate to ecological conditions.

Botkin, in Chapter 9 of this volume, builds on the idea of cultural legacies in aesthetics by exploring key historical transitions in how we have perceived forests and landscapes, as illustrated in art and in the testimony of lay people and practitioners. Botkin suggests that the landscape painters of the 17th and 18th centuries helped to shape modern attitudes to landscape, through their depiction of a somewhat idealized nature. He also warns against the influence of predisposition's based on inadequate scientific evidence, and argues for deeper ecological understanding in formulating opinions on the landscape.

Bell further examines several historical theories on aesthetics in Chapter 10 beginning with the senses and their use in perception and then moves on to consider Marr's primal sketch theory, Gestalt psychology and Gibson's ecological theory of perception; from this, an understanding of landscape pattern recognition is presented as a fundamental basis for the aesthetic experience. In the second part of Chapter 10, Bell explores the nature of landscape aesthetics, particularly the role that knowledge about the perceived scene plays, and the exact nature of an aesthetic experience as presented by Foster. A review of the ideas of Leopold and Gobster considers the manner in which aesthetics and ecology interact; Bell concludes that an understanding of preferences alone is insufficient to guide managers.

The exploration of various aesthetic theories and the potential relationships (or lack of relationships) between visual quality and indicators of forest sustainability, raises a number of questions for Sheppard in Chapter 11. He reviews the contribution of visual resource management (VRM), as connected by agencies such as in the US Forest Service and the British Columbia Ministry of Forests, in promoting a *scenic aesthetic*. He contrasts this with the emerging *ecological aesthetic* (as discussed in Gobster's Foreword), and proposes a refined theory of visible stewardship, as a partial solution to adverse public reactions to forestry management practices that ecologists might wish to promote.

From a social perspective, landscapes encompass far more than ecological and aesthetic values. Kruger, in Chapter 12, posits that there is more to landscapes than meets the eye, through an exploration of landscapes as places with which people develop relationships and to which they ascribe meaning. This chapter agues that civic engagement of a variety of stakeholders in social learning, stewardship activities and other collaborative processes are needed to explore options and potential outcomes successfully. Kruger argues that the development of frameworks is also needed to bring together local knowledge of places with scientific knowledge to help inform decision-making and to provide for the assurance of broad-based sustainability.

5   Part V: Visualization of Forested Landscapes

At the same time that our awareness of sustainability issues has been developing, there has been rapid growth in our ability to understand ecosystem processes and to make them known to people using a variety of media. The role of image-making by painters from the 17th to the 20th century (as discussed by Botkin in Chapter 9) has been to some extent replaced by landscape photography as a means of revealing attitudes to the ecological health of the environment: Walker-Evans' evocative images of the dust-bowl disaster of the 1930s, Ansel Adams' and David Muench's celebration of natural landscape beauty, and Peter Jennings' "God's own Junkyard" compilation spring to mind. The work of photographers such as these has now in turn been complemented by a new wave of powerful, analytically-oriented imagery. Just as the naturalists' descriptive studies of nature have been replaced by the scientists' quantitative data and dynamic modelling, so the static photograph has been superseded by computer-generated scientific visualization and interactive visual simulations of ecosystem processes. The potential contributions of these parallel advances in technology, in bringing both scientific knowledge and public sensibilities together on natural resource decision-making, form the subjects addressed in *Part V*. This section also explores the potential role of landscape visualizations in improving our understanding of how people perceive sustainability, and charts some of the limitations and concerns raised by the use of these media.

Can the employment of visual simulations help to understand and interpret public preferences for sustainable landscapes? The answers to this question depends upon our understanding of aesthetic preferences and our ability to accurately depict and interpret sustainable landscapes. Luymes, in Chapter 13, suggests that advances in visualization technology which now allow forest managers to portray realistic-appearing outcomes, will increase both their use and the dependence on them by forest managers to get their point across to the public. The increased importance of visual simulations, however, may have both positive and negative consequences, including the likelihood that they will carry with them an assumed authority not necessarily supported by scientific rigour or objectivity. Luymes argues that these problems may be most acute in the more sophisticated visual simulations, highlighting the risks and potential of these powerful tools when used with explicit rhetorical intent. In particular, he considers the role of visualization in exploring the spatial and visible forms of new (or ecological) forest management practices, and shaping (rather than merely measuring) the public's aesthetic preference for landscapes.

The potential for the interactivity that visualization and simulation can allow offers opportunities to engage the public in meaningful forest management participation; however the full potential of these opportunities has not been realized. Orland and Uusitalo, in Chapter 14, examine in more depth the suitability of virtual reality in supporting forest-land managers in their involvement with the public in forest planning. One of the leading examples of state-of-the-art visualization tools that aid forest management, *SmartForest*, is examined and evaluated using a case study.

If 17th and 18th century landscape painters were able to interpret and influence the scenic aesthetic using technologies of the day, can we assume that modern visualization and modelling techniques can do the same for the development and refinement of an ecological aesthetic? Danahy, in Chapter 15, returns to this theme, and suggests that the most important factor governing the selection of a visualization technique is the identification of the specific aesthetic and ecological meanings one is attempting to communicate or replicate in a simulation. The techniques available for visualization provide a diversity of high quality representations, yet the true potential of these techniques lies in properly matching them to our capacity to systematically associate abstract knowledge with visually explicit representations: in short, do people have the necessary visual literacy to make the appropriate connections between the simulated image and the ecological reality?

Wherrett, in Chapter 16, describes the use of psychophysical landscape preference models (as outlined by Daniel in Chapter 2) for the prediction of scenic beauty and notes that few models have been validated and fewer still used in a real landscape planning context. She describes an experiment that applies a landscape preference model using a selection of landscape visualization techniques, and assesses the realism and validity of the visualizations themselves, as an example of some of the issues mentioned by the preceding authors in *Part V*.

6   Part VI: Reconciling Forest Sustainability and Aesthetics

The volume concludes with a synthesis of key conclusions and unresolved dilemmas, drawn from the individual chapters, from additional points raised at the PWIAS Exploratory Workshop, from the combination of authors' views, and from previous research. It presents an initial research agenda to address the fascinating and under-explored area of sustainability, ecology, and aesthetics, as well as some implications for forest resource managers.

It is hoped that this book will catalyse academic, practitioner, and lay person curiosities on the subject of linking sustainability to aesthetics, and encourage debate among and between the many disciplines involved in the subject areas of forestry, ecology, socio-economics, philosophy, psychology, landscape architecture, and computer science. Further, we hope that the research agenda developed here can form a platform for new research to clarify further the relationships between forest sustainability and people's aesthetic responses.

References

British Columbia Ministry of Forests (1997) *Managing Visual Resources to Mitigate Timber Supply Impacts*. Draft Memorandum from Henry Benskin, Director Forest Practices Branch, to Larry Pederson, Chief Forester. Forest Practices Branch, BC Ministry of Forests, Victoria, BC.

Leopold, A. (1968) *A Sand County Almanac and Sketches Here and There*. Oxford University Press, London.

Peter Wall Exploratory Workshop (1999) *Linking Aesthetics to Sustainability: Do People Prefer Sustainable Landscapes?* University of British Columbia, Vancouver, 23-27 February 1999. <http://www.forestry.ubc.ca/pwall/default.htm>.

PART II

Seeing and Knowing: Approaches to Aesthetics and Sustainability

Chapter two:

Aesthetic Preference and Ecological Sustainability

Terry C. Daniel

*Department of Psychology, Environmental Perception Laboratory, University of
Arizona*
Tucson, Arizona

1   Do People Prefer Sustainable Landscapes?

This question was posed to the participants in the Peter Wall Institute for Advanced
Studies (PWIAS) Workshop. In many respects the question was intended as a
straw man to stimulate discussion of a wide range of philosophical, biological and
psychological/social questions relevant to environmental policy and management.
In that regard the question was very successful. Participants vigorously debated the
relationships between natural/ecological values and human values, with particular
attention to landscape aesthetics. The goal of this chapter is to address the question
from a narrower, empirical perspective. That is, to assume (perhaps naively)
that part of the motivation for posing the question, and one of the goals for the
workshop, is to seek an answer, or at least to frame the conditions for finding an
answer.

On the surface the question would appear to be rather straightforward, offering
the prospect of a direct empirical answer. One need only find a number of landscapes
that do and do not exhibit *sustainability*, present them to people and determine
which they *prefer*, or empirically, which they choose. That is, to determine
the relationship between bio-physical features of landscapes (sustainability) and
human psychological response (preference/choice), a psychophysical relationship.
This rendering of the question may be faulted on a number of counts. It may raise
as many difficult operational questions in the biological and psychological domains
as it avoids in the philosophical domain. On the other hand, a psychophysical
interpretation does have some important advantages. First, the question becomes,
at least in principle, answerable. Second, there is a substantial and mature research
literature addressing human landscape preferences, including quantitative methods
for specifying psychophysical relationships between bio-physical landscape
features and landscape preferences (Daniel, 1990). But taking this direct approach
requires an agreed upon operational means of distinguishing sustainable from

unsustainable landscapes and the specification of an appropriate meaning for preference.

2 Sustainability

Many different definitions for sustainability have been offered, in some cases distinguishing biological sustainability from related constructs, such as *social sustainability* or *sustainable development*, but often not. Some bureaucratic definitions of sustainability would seem to make the answer to the workshop question a foregone conclusion. The USDA Forest Service, for example, has defined sustainable forest management as "The management of forest and rangelands for environmental protection and economic stability..." (USDA Forest Service, 2000). By this definition, it would seem very unlikely that anyone would not prefer sustainable over unsustainable landscapes. If the sustainability referenced in the workshop question is to include both environmental protection *and* economic benefits, the question would pit the combination of biological/ecological and most other environmental resource values associated with sustainable landscapes against those same values in an unsustainable form. Who would not want environmental protection and economic benefits for as long as possible? But this is not likely to be the intent of the workshop question.

The workshop question more likely revolves around actual (or assumed) conflicts between human values/desires/needs and biological/ecological values. The motivation for raising such a question is the belief by many that environmental management policies directed at providing benefits to people are leading to the destruction of natural biological processes and ecosystems. In this context some have argued that biological/ecological values are *intrinsic*, and require protection in their own right, independent of any human wants and needs (*e.g.* Callicott, 1983; 1985; Rolston, 1984; 1988). This approach tends to make the question a contest between two fundamentally different types of values, *nature values* and *human values*. Alternative arguments (*e.g.* James, 1891; Murdy, 1975; Weston, 1985; Seligman, 1989; Harlow, 1992) insist that the relevant values (indeed all values) must be anthropocentric. But this does not preclude a preference for ecologically sustainable landscapes based on the value of providing for future human wants and needs - on instrumental rather than intrinsic values of the ecosystem (*e.g.* Daniel, 1988; Parsons *et al.,* 1993). Preferences for (or against) ecologically sustainable landscapes, by an instrumental account, would not be construed in terms of moral/ethical obligations toward ecosystems/nature, but in terms of the level of contemporary human concern for the welfare of future generations of humans.

Thus, the question posed addresses the relationship between *ecological* sustainability and human landscape preference. It is outside the scope of this chapter to attempt to resolve the substantial philosophical issues regarding whether preferences for ecological sustainability should be supported on the basis of intrinsic or instrumental values. The specification of the question sought here is not directed at resolving ethical, or even normative issues of what "should" be preferred or what the basis of such a preference ought to be. Both intrinsic and instrumental value orientations leave open the question of whether people do or do not in fact prefer ecologically sustainable landscapes. But what does "prefer" mean in this context?

3 Preference

Much of the existing psychological research in human preference has been based on a "rational" model (*e.g.* Edwards, 1954; Simon, 1955; von Winterfeldt and Edwards, 1986). This approach places considerable emphasis on logical, computational processes in choice and decision making (*e.g.* Reed, 1982; Payne *et al.*, 1992). Applying the rational model to ecologically sustainable landscapes in particular, preferences should be strongly influenced by ecological knowledge. For example, people with greater knowledge of ecosystems should be more likely to prefer ecologically sustainable landscapes (*e.g.* Carlson, 1977; 1995; Gobster, 1999). Alternatively, recent preference research and theory has placed (renewed) emphasis on the role of "affective" (as distinguished from "cognitive") processes (*e.g.* following Zajonc, 1980). Evidence is growing that logical/rational processes (exemplified by language) and emotional processes may be carried out in separate brain systems significantly insulated from each other (*e.g.* Buck, 1985; Damasio, 1994; LeDoux, 1995). Environmental psychologists have long acknowledged that affect (emotion) plays a major role in determining human response to environments (*e.g.* Ittelson, 1973; Wohlwill, 1976). Research has confirmed that views of environments (landscapes) produce strong effects on important emotion-related psycho-physiological responses (*e.g.* Ulrich, 1981; Hartig *et al.*, 1991; Parsons, 1991; Ulrich *et al.*, 1991). Moreover, there is some indication that emotional responses to environments may be fundamentally different from those elicited by social stimuli (Parsons *et al.*, 1998). In short, environmental preferences may depend more on specialized affective reactions than on any knowledge-based logical operations.

Consistent with the noted role of affective processes, environmental (and many other) preferences have come to be viewed as having a substantial genetic/ evolutionary basis (*e.g.* Ulrich, 1983; Kaplan, 1987; Kaplan and Kaplan, 1989; Parsons, 1991; Orians and Heerwagon, 1992; Ulrich, 1993). That is, important keys to understanding contemporary human preferences are to be found in emotional adaptations developed through natural selection during human evolution (*e.g.* Tooby and Cosmides, 1990). While the affective/evolutionary model questions the role of contemporary knowledge and reason, it does not necessarily argue against a contemporary human preference for ecologically sustainable landscapes. The intimate interaction with the natural environment that characterized human evolution would insure some correspondence between human preferences and natural ecological conditions that were sufficiently stable to affect the natural selection/adaptation process. On the other hand, neither does this model insure that contemporary human landscape preferences would be positively related to contemporary conceptions of ecological sustainability. It is not clear, for example, how natural selection could make the 30-something human cycle particularly responsive to 1000-something ecological cycles. The pre-historical record suggests that when environmental conditions became significantly unfavourable to humans, the humans either died, moved on or, within the limits of their technology, changed the environment (Glacken, 1967; Dubos, 1972; 1980). Thus, the affective/ evolutionary model of environmental preference, like the rational choice model before it, leaves open the question of whether ecologically sustainable landscapes are in fact preferred.

4   Ecological Preferences

There are reasons to suspect that, while the public may generally support the concept of ecosystem management, they may not like many of the specific effects. Many of the stages of natural succession in forest ecosystems exhibit features that are unattractive to people and that are not conducive to the most favoured forest recreation activities. The natural change agents in forest ecosystems (*e.g.* insects, disease, fire, wind, and decay) have not typically been appreciated by the public either as processes or for their most obvious and visible effects on the forest. Management efforts to *restore* or mimic these natural processes may be even less palatable.

It can be argued that on the ecological time-scale many of the *unattractive* conditions in the forest are only transitory, and that these conditions are often prerequisites for later conditions that are attractive to the public. Moreover, the forest conditions that the public most desires are themselves often only transitory, and in the long term impossible to maintain. But ecologically transitory conditions can prevail for several human generations in some forests. The fact that temporary undesirable states of the forest may herald potentially much more desirable conditions in the ecological future may be judged as offering little consolation to current visitors (or their children or even grandchildren). Similarly, unattractive conditions do not typically prevail everywhere in the forest. There will always be attractive conditions somewhere. The forest must be thought of as a mosaic of conditions, a dynamic mosaic that naturally and continually changes in response to ecosystem processes. But this geographic variability in natural ecosystems may not be especially appreciated by the visiting/residing public, especially if the attractive conditions are not in the locations that the public visits, or if the areas of unattractive conditions happen to be "in my backyard."

5   Preferential Consensus

Whether viewed as predominantly a cognitive or an affective process, preference is not a *stand alone* construct. In addition to an object (something to be preferred) it requires a subject (agent) to do the preferring, and some purpose for doing so. In the present context the specified object of preference is a landscape, exhibiting yet to be fully identified features relevant to ecological sustainability. Background material and much of the discussion that ensued at the PWIAS workshop indicates that the intended subject (agent) is one or more of the various segments of "the public" that are usually identified as *stakeholders* in environmental planning and policy formulation contexts. To the extent that landscape preferences are subjective ("in the eye of the beholder"), the answer to the workshop question might well differ depending upon which segment of the public's preferences are being assessed. Lack of consensus does not necessarily present a major problem, however, as it is routine in social science-based preference assessments (*e.g.* public opinion polls, market surveys) to distinguish among sub-sets of the more general population when statistical analysis reveals the need.

People, of course, will differ in their aesthetic preferences (as in almost every other aspect). The relevant question, however, is whether differences in individual (or sub-group) preferences are large or small relative to the differences in preference between sustainable and unsustainable landscapes. Is the within-publics variance greater than the between-landscapes variance? Three decades of landscape assessment research overwhelmingly shows that preferences for natural or near-natural landscapes indicated by choices, rankings or ratings of *like-dislike*, *visual quality*, *natural scenic beauty*, or *preference* are highly consistent (*e.g.* Rabinowitz and Coughlin, 1971; Daniel *et al.*, 1973; Zube *et al.*, 1974; Daniel and Boster, 1976; Craik and Zube, 1977; Shafer and Brush, 1977; Buhyoff and Leushner, 1978; Ribe, 1990). Between-landscapes variance is typically found to be several orders of magnitude greater than the variation in preferences between individuals (the landscape-by-observers interaction), even in studies that are designed to make cross-cultural comparisons (*e.g.* Shafer and Richards, 1974; Zube, 1974; Daniel and Boster, 1976; Ribe, 1994; Sommer and Summit, 1996). Thus, it is very likely that any reasonably representative sample would be sufficient to determine the central tendency in the relationship between people's landscape preferences and ecological sustainability. Individual or sub-group differences should not be assumed *a priori*, but if empirical evidence of systematic differences in patterns of landscape preferences is found, appropriate qualifications in the specification of relationships should be made explicit.

6   Purposes for Preferences

As noted above, the question posed to the workshop was intended in part as a straw man for the human versus ecological/natural values conflict currently receiving great attention in environmental philosophy and management/policy. In this context, the purpose for the preference referred to in the question might be construed to be something like "to achieve the best of all possible worlds." Such a purpose provides fertile ground for philosophical debates and for wide-ranging and important discourse on social policies and environmental ethics - an unquestionably worthwhile cause. But such a broad construal of the purpose for preference would virtually prohibit any empirical approach to answering the question.

There is evidence in the statement of the question that such an all-encompassing purpose was not intended. For example, the question specifically identifies the preference object as a *landscape*, rather than an environment or an ecosystem or a *state of the world*. The term landscape has undergone considerable expansion in recent years, so that it has tended to encompass virtually all aspects of the environment (*e.g.* Daniel and Vining, 1983; Daniel, in press). For purposes of this chapter, and consistent with most dictionary definitions and the intentions implied in the background and discussions associated with the PWIAS Workshop, landscape will be taken to refer to the *visible* land surface over some reasonably (at least in principle) viewable area. Thus, the question is taken to refer to preferences based on the visible features of the environment.

Visual inspection of the landscape could provide a basis for a number of preference purposes. For example, one might be able to estimate the preferability (suitability) of a site for residential or other development, for one or another recreational activity, or for agricultural uses. Indeed, visual inspection might allow reasonable estimates of the quality of biological/ecological parameters, such as relative levels of biomass, vigour, biodiversity or even ecological sustainability. But all of these would be estimates, which might have to be modified based on other, non-visual information. Neither the utilitarian nor the ecological examples cited are fundamentally visual - there are other, non-visual criteria that are more essential. For example, a site may visually appear to be highly preferred for home construction, but then be found to have subsurface characteristics (*e.g.* hard rock or high levels of radon gas) that make the site very unsuitable. Similarly, a site that appears to be highly preferable for biodiversity purposes may prove to be much less so based on more direct biological inventory and analysis.

For purposes such as those cited above it is not surprising, and it may indeed be expected that visual assessments do not correlate perfectly with assessments based on non-visual (*e.g.* physical or biological) criteria. That is, visually-based judgements could be in error. But the workshop question is not about the relationship between visually based estimates of biological qualities (ecological sustainability) versus biologically-based assessments. Rather, the preference at issue has a more fundamentally visual basis that is at least potentially independent of biological parameters. Thus, it would not be appropriate to view a preference for landscapes that prove on biological grounds to be ecologically unsustainable to be in "error" (though some might find such a preference ethically wrong). The workshop question was generally agreed to be about *aesthetic* preferences,

preferences that do not rely in any necessary way on biological/ecological factors, or on economic/utilitarian considerations.

Just what is meant by the term "aesthetic" is a long-standing, and controversial topic. Two philosophical positions that have been especially influential in landscape aesthetics research and practice are the formal/objective, in which aesthetics is viewed as a property of the aesthetic object, and the subjective, in which aesthetics is viewed as a particular type of human response to/experience of an object (Lothian, 1999). Subjective models have generally associated aesthetics with judgements of *beauty*, or expressions of *liking* or *preference* in contexts in which any specific utilitarian motives are set aside (Zube *et al.*, 1982; Daniel and Vining, 1983). In the context of the workshop question, the subjective model places visual aesthetic landscape preferences in the human viewer, and distinguishes aesthetic purposes from utilitarian or biological purposes.

7 Bio-Philosophical Issues

Narrowing of the workshop question to the psychophysical relationship between ecological sustainability and human visual aesthetic preferences enables in principle an empirical approach to answering the question. By what may be a bit out of fashion standards this specification renders the question *scientifically meaningful* (see the opposing views by Bengston, 1994 and Hetherington *et al.*, 1994). Still, there remain a number of significant obstacles to finding an answer. Perhaps most important is the lack of any clear, consistent and agreed upon way to distinguish landscapes systematically on the basis of ecological sustainability. Substantial questions remain regarding the appropriate biological definition/criteria for ecological sustainability, the meaning of which has yet to reach a high level of consensus within the professional ecology/environmental science and management community. Moreover, it seems unlikely that environmental conditions that promote sustainability in one ecosystem will be the same as sustainability conditions in another ecosystem. Thus, sustainable conditions in one ecosystem might be highly preferred, while sustainable conditions in another ecosystem might not be preferred. A significant obstacle to answering the workshop's primary question, then, is the inability of current biological/ecological science unambiguously to identify instances of ecologically sustainable and unsustainable (or less sustainable) landscapes, within and between different ecosystems. This problem will not be further addressed here, but is left with an urgent plea for attention by the relevant biological and ecological disciplines.

The psychophysical interpretation of the workshop question admittedly side steps some substantial philosophical issues regarding the proper role of environmental ethics, and the multifaceted issues in the philosophy and psychology of environmental aesthetics. These issues will not be further addressed in this chapter. In defence, some of these issues have received more concerted attention in previous (Daniel, 1988; Hetherington *et al.*, 1994; Parsons *et al.*, 1993) and forthcoming papers (Daniel, in press). In spite of these admitted shortcomings, the psychophysical approach does allow, in principle, for people individually or collectively to prefer (or not) landscape conditions that are (or are not) ecologically sustainable. The remaining sections of this chapter will be devoted to what are largely operational/methodological issues in the psychological domain that impede an empirical determination of the psychophysical relationship between ecological sustainability and visual aesthetic landscape preferences.

8   Operational Challenges

Preference (*visual quality, scenic beauty*) has most often operationally been construed as a static concept in landscape quality assessment research. A particular set of landscape conditions (a landscape scene) is typically presented on a single occasion and observers indicate whether they prefer that set of conditions more or less than other sets of conditions (other scenes). But this approach to assessing landscape aesthetic preference is not sufficient to address the question of whether ecologically sustainable landscapes are preferred. Sustainability is necessarily a temporally and geographically dynamic concept. An ecologically sustainable landscape is most likely to be characterized by a spatial mosaic of environmental conditions that changes in some (natural) progression over time. The geographic scale may be very small, very large or mixed, and changes may be very gradual or abrupt, minimally or catastrophically disturbing, depending upon the ecosystem. Thus, to prefer an ecologically sustainable landscape must mean to prefer a particular geographic and temporal pattern of environmental conditions over other possible patterns. Such preferences cannot adequately be assessed by presenting individual scenes depicting particular places at particular points in time.

Even when restricted to visual aesthetics, determining preferences for geographically and temporally varying patterns of landscape conditions poses substantial challenges to psychological/perceptual assessment methods. What are the necessary and sufficient requirements for representing geo-temporal landscape patterns to observers so that expressed preferences will be valid?

If the relevant range of spatial variations can be covered within a single view (a vista), individual scenes depicting sustainable and unsustainable patterns might suffice. But this method would not capture the effects of temporal variations in the geographic pattern. The application of appropriate computer visualization technology (Sheppard, 1989; Vining and Orland, 1989; Bishop and Hull, 1991; Daniel, 1992; Orland, 1993; Daniel and Meitner, in press) would allow the presentation of temporal sequences of scenes or *time-lapse* movies showing changing spatial patterns in each vista. But, would expressed preferences based on such presentations be valid? In many ecosystems the rate of change in landscape patterns can be very slow - involving intervals that substantially exceed a human lifetime. It is technically possible to compress these changes and present them in a few minutes. But are people's expressed aesthetic preferences valid or even meaningful in such a context? Psychological theories of preference offer no obvious answer, and there would not appear to be any direct way to answer this question empirically.

Many environments in which ecological sustainability and human aesthetic preferences are potentially relevant do not naturally provide views that would contain even a small part of the range of geographic variations. For example relatively flat landscapes with dense forest vegetation do not offer vista perspectives. One representation option here might be aerial views that do capture the relevant range of variation. The temporal variation issue discussed above would still be obtained, but added to that would be the unnatural viewpoint. For example, the natural experience might be of restricted views penetrating through tree trunks and vegetation underneath the canopy of the forest. Aerial perspectives change this experience to expansive views looking down on the top of the canopy of the forest, with no view of trunks or vegetation underneath. While the technology is clearly available to provide observers with *fly-overs* and even to show temporal changes in aerial views with time-lapse techniques, it is not clear that such representations would support valid assessments of landscape aesthetic preferences.

A more natural way to experience geographical variations in restricted-view landscapes is to move through them, encountering different ecological conditions sequentially as they occur. Computer visualization technology could readily provide such *walk-through* visual representations and some important presentation parameters are reasonably well established. Virtual reality techniques could provide the observer with realistic visual experiences, including freedom to look about the landscape and to proceed into it in any chosen direction. As with the vista perspective discussed above, the temporal variations in landscape conditions that are a central component of ecological sustainability introduce additional issues that have not been addressed in the psychology of preference or in visualization science. In addition to questions about the appropriate rate of change to present, it must be determined whether changes should be depicted place-by-place as they are encountered within a given virtual trip through the landscape, or trip-by-trip, with separate trips being completed at each relevant time step. Again preference theories provide little direction for specifying the appropriate procedures, and there is no direct way to resolve these issues empirically.

It is clear that the *traditional* method of presenting single scenes representing conditions at a particular place and point in time is not a sufficient procedure for determining people's preferences for ecologically sustainable landscapes. What is less clear is what the necessary and sufficient representation/presentation procedures would be. Perhaps the best approach would be to apply several different procedures (various versions of time-lapse walk-throughs and/or fly-overs, for example), to use converging operations in an attempt to *surround the truth*. To the extent that ecologically sustainable landscapes were aesthetically preferred over unsustainable landscapes across multiple assessment methods, one could relatively safely conclude that people do in fact prefer sustainable landscapes. But what if unsustainable landscapes were the most preferred? Should people's aesthetic preferences be changed? To what? By whom?

9 Political Dilemmas

The psychological literature, including psychotherapy, persuasive communication, marketing and advertising provides some methods that might be effective in changing public aesthetic preferences. But it is not clear what individuals, institutions, or agencies have the political mission, or the moral licence to determine whether and/or what changes are appropriate. There is continuing, legitimate debate about the extent to which environmental management agencies should "give the public what they want" versus following what agency experts judge is "best for the environment" (which might arguably be in the long-term best interest of the public). But it may be dangerous in a democratic society for government agencies to become too actively involved in shaping public opinion and values. If changing landscape aesthetic preferences only required providing *education* about ecological processes (changing knowledge), this might not pose a particularly difficult political problem. However, if more *forceful* procedures were required (*e.g.* procedures that could be construed as propaganda or conditioning), as might well be the case if evolution-based emotional responses are central to landscape aesthetic preferences, the political issues could become much more salient, and thorny.

10 Conclusion

Do people prefer sustainable landscapes? Answering this question requires refining the meanings of both sustainable landscapes and preference. Limiting the question to the psychophysical relationship between ecological sustainability and visual aesthetic preference yields a scientifically meaningful and potentially answerable question. The meaning of ecological sustainability must be biologically determined - a task that remains for the relevant biological and ecological disciplines. Psychophysical methods for determining human preferences are well developed, but determining preferences for ecologically sustainable landscapes presents special operational problems that have not been resolved in theory or practice. Sustainable landscapes will likely be characterized by mosaics of environmental conditions that vary over both space and time. To prefer sustainable landscapes is to prefer particular geo-temporal patterns of landscape conditions, but the scale of these patterns often exceeds the human scale. It is as yet unclear how sustainable and unsustainable patterns (once biologically defined) should be presented to human observers to obtain valid expressions of aesthetic preferences. In the absence of a clearly correct method, a converging operations approach is advocated as a means of finding any consistent, method-independent landscape preferences.

If the answer to the workshop question is that people prefer unsustainable landscapes, potentially difficult political questions emerge. Should public aesthetic preferences be changed to favour sustainable landscapes? How should such a change be effected, and by whom? Changing aesthetic preferences could require manipulating some or all of human perceptions, thoughts and feelings about the landscape. Current psychological research and theory emphasizes the *modularity* of psychological processes, implying that changing human aesthetic preferences could require very different operations depending upon whether perceptions, cognitions or emotions were the appropriate target. One can not rule out the possibility that preference for ecologically sustainable landscapes could conflict with human nature, *i.e.*, with evolved affective responses.

While the data regarding environmental aesthetic preferences in particular are limited, preferences more generally are notorious for their resistance to *cognitive* manipulation ("preferences need no inferences"). As an often cited example, providing knowledge of the nutritional benefits of spinach will have little effect on the (aesthetic) preferences of one who hates the taste of the stuff. To the extent landscape aesthetic preferences are like food (and many other) preferences, it is unlikely that operations focused on changing knowledge about ecological processes alone would bring about any substantial changes. Ecological knowledge is probably not sufficient to produce an aesthetic preference for sustainable landscapes. On the other hand, methods that have shown some ability to alter basic emotional responses, such as classical conditioning, some forms of psychotherapy and strong persuasive communications techniques, may not be socially, politically or morally acceptable means. What then could be done about public preferences for unsustainable landscapes?

Environmental philosophers, deep ecologists, journalists and concerned citizens (individually and as special interest groups) have legitimately argued for policies that promote ecologically sustainable landscapes, and have lobbied public support for such policies. In some instances the ecological agenda has been pursued in courts and advocated in public demonstrations. These are all tried and accepted methods for effecting policy changes in democratic societies. The role that public agencies should play in promoting ecological sustainability as public environmental management policy is more controversial, especially when affecting changes in public preferences requires (as it almost certainly will) more than providing relatively *cold* information about the pros and cons of alternative policies. Thus, if people do not prefer sustainable landscapes (which has yet to be determined) public environmental management agencies convinced of the need to manage for ecological sustainability would be faced with a particularly difficult dilemma. A possible resolution could be to settle for less than a change in public aesthetic preferences. Perhaps it would be sufficient for people to accept ecologically sustainable landscapes as one important goal for environmental management policy, even if they do not find such landscapes to be aesthetically preferred. If people were convinced of the nutritional benefits, they might well be willing to eat their spinach, even if they still do not especially like the taste.

References

Bengston, D.N. (1994) Changing forest values and ecosystem management. *Society and Natural Resources*, 7: 515-533.

Bishop, I.D. and R.B. Hull (1991) Integrating technologies for visual resource management. *Journal of Environmental Management*, 32: 295-312.

Brush, R.O. (1979) The attractiveness of woodlands: Perceptions of forest landowners in Massachusetts. *Forest Science*, 25(3):495-506.

Buck, R. (1985) Prime theory: an integrated view of motivation and emotion. *Psychological Review*, 92: 389-413.

Buhyoff, G.J., and W.A. Leuschner (1978) Estimating psychological disutility from damaged forest stands. *Forest Science*, 24: 424-432.

Callicott, J.B. (1983) Leopold's land aesthetic. *Journal of Soil and Water Conservation*, 38: 329-332.

Callicott, J.B. (1985) Intrinsic value, quantum theory and environmental ethics. *Environmental Ethics*, 7: 257-275.

Carlson, A.A. (1977) On the possibility of quantifying scenic beauty. *Landscape Planning*, 4: 131-171.

Carlson, A.A. (1995) Nature, aesthetic appreciation and knowledge. *The Journal of Aesthetics and Art Criticism*, 53: 394-400.

Craik, K. and Zube, E. (eds) (1977) *Perceived Environmental Quality Indices*, Plenum, New York.

Damasio, A.R. (1994) *Descarte's Error: Emotion, Reason and the Human Brain*. G.P. Putnam's Sons, New York.

Daniel, T. C. (1988) Human values and natural systems: a psychologist responds. *Society and Natural Resources*, 1: 285-290.

Daniel, T.C. (1990) Measuring the quality of the human environment: a psychophysical approach. *American Psychologist*, 45: 633-637.

Daniel, T.C. (1992) Data visualization for decision support in environmental management. *Landscape and Urban Planning*, 21: 261-263.

Daniel, T.C. (in press) Whither scenic beauty? Visual landscape quality assessment in the 21st Century. *Landscape and Urban Planning*.

Daniel, T.C. and R.S. Boster (1976) *Measuring Landscape Esthetics: The Scenic Beauty Estimation Method*, USDA Forest Service Research Paper, 167, Ft. Collins, CO.

Daniel, T.C. and M.J. Meitner (in press) Representational validity of landscape visualizations: the effects of graphical realism on perceived scenic beauty of forest vistas. *Journal of Environmental Psychology*.

Daniel, T.C. and J. Vining (1983) Methodological issues in the assessment of landscape quality. In: I. Altman and J. Wohlwill (eds), *Human Behavior and Environment, Vol. VI*. Plenum, New York.

Daniel, T.C., L. Wheeler, R.S. Boster and P. Best (1973) Quantitative evaluation of landscapes: an application of signal detection analysis to forest management alterations. *Man-Environment Systems*, 3: 330-344.

Dubos, R.J. (1972) *A God Within*. Angus and Robertson, London.

Dubos, R.J. (1980) *The Wooing of Earth*. Charles Scribner's Sons, New York.

Edwards, W. (1954) The theory of decision making. *Psychological Bulletin*, 51: 380-417.

Glacken, C.J. (1967) *Traces on the Rhodian Shore: Nature and Culture in Western Thought from Ancient Time to the End of the Eighteenth Century*. University of California Press, Berkeley.

Gobster, P.H. (1999) An ecological aesthetic for forest landscape management. *Landscape Journal*, 18(1): 54-64.

Harlow, E.M. (1992) The human face of nature: environmental values and the limits of anthropocentrism. *Environmental Ethics*, 14: 27-42.

Hartig, T., M. Mang and G.W. Evans (1991) Restorative effects of natural environment experiences. *Environment and Behavior*, 23(1): 3-26.

Hetherington, J., T.C. Daniel and T.C. Brown (1994) Anything goes means everything stays: The perils of uncritical pluralism in the study of ecosystem values. *Society and Natural Resources*, 7(6): 545-546.

Ittelson, W.H. (1973) *Environment and Cognition*. Seminar, New York.

James, W. (1891) The moral philosopher and the moral life. International Journal of Ethics, April, 1891. Reprinted in: A. Castell (ed) *Essays in Pragmatism by William James, 1948*. Hafner Publishing Co., New York, pp. 65-87.

Kaplan, S. (1987) Aesthetics, affect, and cognition: Environmental preference from an evolutionary perspective. *Environment and Behavior*, 19(1): 3-32.

Kaplan, R. and S. Kaplan (1989) *The Experience of Nature: A Psychological Perspective*. Cambridge University, Cambridge.

LeDoux, J.E. (1995) Emotion: clues from the brain. *Annual Review of Psychology*, 46: 209-235.

Lothian, A. (1999) Landscape and the philosophy of aesthetics: is landscape quality inherent in the landscape or in the eye of the beholder? *Landscape and Urban Planning*, 44: 177-198.

Murdy, W.H. (1975) Anthropocentrism: a modern version. *Science*, 187, 1168-1172.

Orians, G.H. and J.H. Heerwagon (1992) Evolved response to landscapes. In: J.H. Barkow, L. Cosmides and J. Tooby (eds) *The Adapted Mind: Evolutionary Psychology and the Generation of Culture*. Oxford University Press, New York.

Orland, B. (1993) Synthetic landscapes: a review of video-imaging applications in environmental perception research, planning and design. In: R. Marans and D. Stokols (eds) *Environmental Simulation: Research and Policy Issues*. Plenum Press, New York, pp. 213-251.

Parsons, R. (1991) The potential influences of environmental perception on human health. *Journal of Environmental Psychology*, 11: 1-23.

Parsons, R., T.C. Daniel and L.G. Tassinary (1993) Landscape aesthetics, ecology and human health: in defense of instrumental values. In: W. Covington (ed) *Sustainable Ecological Systems,* USDA Forest Service, Ft. Collins, CO, pp. 266-280.

Parsons, R., L.G. Tassinary, R.S. Ulrich, M.R. Hebl and M. Grossman-Alexander (1998) The view from the road: implications for stress recovery and immunization. *Journal of Environmental Psychology*, 18: 113-140.

Payne, J., J. Bettman and E. Johnson (1992) Behavioral decision research: a constructive processing perspective. *Annual Review of Psychology*, 43: 87-131.

Rabinowitz, C.B. and R.E. Coughlin (1971) *Some Experiments in Quantitative Measurement of Landscape Quality.* Regional Science Research Institute, Discussion Paper Series: No. 43. Philadelphia, PA.

Reed, S.K. (1982) *Cognition: Theory and Applications.* Brooks/Cole Publishing Company, Monterey, CA.

Ribe, R.G. (1990) A general model for understanding the perception of scenic beauty in Northern hardwood forests. *Landscape Journal*, 9: 86-101.

Ribe, R.G. (1994) Scenic beauty perceptions across the ROS spectrum. *Journal of Environmental Management*, 42: 199-221.

Rolston, H. (1984) Are values in nature subjective or objective? In: R. Elliot and A. Gare (eds) *Environmental Philosophy: A Collection of Readings.* University of Queensland Press, New York.

Rolston, H. (1988) Human values and natural systems. *Society and Natural Resources*, 1: 271-284.

Seligman, C. (1989) Environmental ethics. *Journal of Social Issues*, 45(1): 169-184.

Shafer, E.L. and R.O. Brush (1977) How to measure preferences for photographs of natural landscapes. *Landscape Planning*, 4: 237-256.

Shafer, E.L. and T.A. Richards (1974) *A Comparison of Viewer Reactions to Outdoor Scenes and Pictures of Those Scenes.* USDA Forest Service Research Paper NE-302, Northeast Forest Experiment Station, Upper Darby, PA.

Sheppard, S.R.J. (1989) *Visual Simulation: A User's Guide for Architects, Engineers, and Planners.* Van Nostrand Reinhold, New York.

Simon, H. (1955) A behavioral model of rational choice. *Quarterly Journal of Economics*, 69: 99-118.

Slovic, P. (1995) The construction of preference. *American Psychologist*, 50: 364-371.

Sommer, R. and J. Summit (1996) Cross-national rankings of tree shape. *Ecological Psychology*, 8(4): 327-341.

Tooby, J. and L. Cosmides (1990) The past explains the present: emotional adaptations and the structure of ancestral environments. *Ethology and Sociobiology*, 11: 375-424.

Ulrich, R.S (1981) Natural versus urban scenes: Some psychophysiological effects. *Environment and Behavior,* 13(5): 523-56.

Ulrich, R.S. (1983) Aesthetic and affective response to natural environment. In: I. Altman and J. Wohlwill (eds) *Behavior and the Natural Environment.* Plenum Press, New York, pp. 85-125.

Ulrich, R.S. (1993) Biophilia, biophobia, and natural landscapes. In: S.R. Kellert and E.O. Wilson (eds) *The Biophilia Hypothesis*. Island Press, Washington, DC.

Ulrich, R.S., R.F. Simons, B.D. Losito, E. Fiorito, M.A. Miles and M. Zelson (1991). Stress recovery during exposure to natural and urban environments. *Journal of Environmental Psychology*, 11: 201-230.

USDA Forest Service (2000) *USDA Forest Service Strategic Plan (2000)*. <http://www2.srs.fs.fed.us/strategicplan/>

Vining, J. and B. Orland (1989) The video advantage: a comparison of two environmental representation techniques. *Journal of Environmental Management*, 29: 275-283.

von Winterfeldt, D. and W. Edwards (1986) *Decision Analysis and Behavioral Research*. Cambridge University Press, Cambridge.

Weston, A. (1985) Beyond intrinsic value: pragmatism in environmental ethics. *Environmental Ethics*, 7: 321-339.

Wohlwill, J.E. (1976) Environmental aesthetics: the environment as a source of affect. In: I. Altman and J. Wohlwill (eds) *Human Behavior and Environment: Advances in Theories and Research. Vol.1*. Plenum, New York.

Zajonc, R.B. (1980) Feeling and thinking: preferences need no inferences. *American Psychologist,* 35:151-75.

Zube, E.H. (1974) Cross-disciplinary and inter-mode agreement on the description and evaluation of landscape resources. *Environment and Behavior*, 6: 69-89.

Zube, E.H., D.G. Pitt and T.W. Anderson (1974) *Perception and Measurement of Scenic Resources in the Southern Connecticut River Valley* (Pub. No. R-74-1). Institute for Man and His Environment, University of Massachusetts, Amherst, MA, 191 pp.

Zube, E.H., J.L. Sell and J.G. Taylor (1982) Landscape perception: research, application, and theory. *Landscape Planning*, 9: 1-33.

Chapter three:

Aesthetic Preferences for Sustainable Landscapes: Seeing and Knowing

Allen Carlson
Department of Philosophy, University of Alberta
Edmonton, Alberta

1 Introduction

For at least the last thirty-five years there has been in North America a growing awareness of the significance of the aesthetic quality of the environments in which we live, work, and play. Early in this period, attention was focused on what were thought of as *visual resources* with the attendant development of methods of effecting what is still frequently referred to as *visual resource management*. This initial development is exemplified, chronicled, and critiqued in many sources (*e.g.* respectively, Elsner and Smardon, 1979; Zube *et al.,* 1982; Carlson, 1977), and thus need not be further elaborated here. It has since been refined and in part replaced by broader and more sophisticated approaches to conceptualizing and managing the aesthetic quality of landscapes (*e.g.* Kaplan and Kaplan, 1989; Bourassa, 1991; Nassauer, 1997; Bell, 1999; Sheppard, Chapter 11, this volume). However, the visual resource management approach must yet be noted here, for one of its legacies is a continued interest in the preferences of those individuals who are thought of as the consumers of visual resources (Carlson, 1990). Thus, the question of "what kinds of landscapes do people prefer" remains an important topic of discussion.

A more recent trend is a concern for what are typically called *sustainable* environments or landscapes. The core idea of a sustainable landscape is that of one that maintains itself, with or without some human assistance, in a rather viable condition throughout a rather significant period of time (although this, of course, does not imply that sustainable landscapes are totally without change). Such a landscape could thus satisfy the well-known Brundtland sustainable development condition of meeting present needs without compromising the meeting of future needs (Brundtland, 1987). Nonetheless, in spite of the considerable importance of this idea, the required notion of sustainability is difficult to precisely define (Thayer, 1989). Fortunately, however, it is not the purpose of this discussion to define this notion. Rather I simply accept a vague idea of sustainability such as that suggested above and turn to the issue of the relationship, if any, between sustainable landscapes thus understood and the preferences of landscape consumers. Therefore, the question to be considered in this discussion is essentially: Do people prefer sustainable landscapes?

It should be noted, however, that the question of whether or not people actually prefer sustainable landscapes is in the last analysis an empirical question. Thus, since throughout this discussion the focus is on theoretical issues, the empirical matter of what is ultimately the right answer to this question will not be directly addressed. Rather, the discussion will first attempt to sharpen and clarify the question by means of some philosophical analysis, for even to begin to treat any such question, we must first become clearer about what it involves. Second, the discussion will investigate the ways in which, in light of this clarification, people might plausibly be said to prefer sustainable landscapes.

2 A Threefold Distinction

I initiate the process of clarification by drawing a number of distinctions. The first of these is threefold, marking the differences between preferences in general, aesthetic preferences, and aesthetic value. The latter, aesthetic value, should be the ultimate subject of our interest, for in the last analysis it may well be the case that that which has aesthetic value is different from that which some or even most people happen to prefer (Carlson, 1993). Nonetheless, knowledge about people's aesthetic preferences is useful, especially if we wish, for whatever reason, to attempt to accommodate such preferences. Thus, for purposes of this discussion I focus primarily on preferences rather than on value. However, even if we focus only on preferences, it is yet necessary to separate aesthetic preferences from preferences in general. This is primarily because of the obvious point that people's preferences in general may not coincide with their aesthetic preferences. It is frequently the case, for example, that people's preferences in general, being a amalgam of personal, social, moral, economic, and other factors, diverge from and sometimes override their aesthetic preferences. Thus, we must focus explicitly on aesthetic preferences rather than on preferences in general, for to pursue the latter rather than the former might take us far from the topic of this discussion. Our question thus becomes: Do people aesthetically prefer sustainable landscapes?

3 Two Additional Distinctions and Four Cases

Focusing on aesthetic preferences, however, requires understanding what constitutes such preferences. Traditionally and narrowly construed, aesthetic preferences for landscapes have been thought of as preferences for the ways in which landscapes strike our senses, most typically our sense of sight. This is the conception of aesthetic preference that is presupposed in the above mentioned idea of visual resources, and aesthetic preferences thus understood are those that visual resource management aims to satisfy. As noted above, this way of understanding the aesthetic has to some extent been replaced by broader and more sophisticated ideas. However, given our focus on aesthetic *preferences*, it is useful to initially work with this traditional and narrow notion (although we will have reason to move beyond it later). Thus, the question of this discussion, now fully spelled out, becomes: Do people aesthetically prefer the look of sustainable landscapes? Putting the question in terms of the look of a landscape suggests another useful distinction, that between, on the one hand, the case in which two landscapes, one sustainable and one not, have the same (or a very similar) look and, on the other, the case in which they have different looks. This distinction is important, for it might be thought, quite reasonably given the traditional and narrow interpretation of aesthetic, that aesthetic preferences will differ only when landscapes look different, quite independent of whether or not they are sustainable. I return to this issue in Section 6.

In addition to whether or not two landscapes look the same or look different, however, the reality behind the appearances of different landscapes is also important to keep in mind. In line with this, yet another significant distinction must be noted. This is the distinction between, on the one hand, our knowing (or our having reasonable and strong beliefs) about the sustainability of a landscape, and, on the other, our not knowing (or not having any beliefs) about it. Thus, if we consider the possibility of two landscapes, one sustainable and one not, as well as our knowledge of (or our beliefs about) the sustainability of these two landscapes, together with the previously introduced distinction between such landscapes either having the same look or having different looks, the upshot is four different possibilities: (1) the two landscapes look the same and we do not know which is which, (2) they look different and we do not know which is which, (3) they look the same and we know which is which, and (4) they look different and we know which is which. It is important to distinguish these four different possibilities, for our focus question about people's aesthetic preferences for sustainable landscapes might be answered differently according to which case is being considered.

4   Cases 1, 2 and 3

The best illustration of clearly different answers to our focus question depending on which of the above indicated possibilities is considered is *Case 1*, which seemingly differs dramatically from all of the other three cases. This is because it may plausibly be argued, concerning *Case 1*, that there are no grounds whatsoever for any differences in preferences, aesthetic or otherwise. If we confront two different landscapes that look the same and, even though one is sustainable and the other not, we have no knowledge (or beliefs) about which is which, then there would seem to be no possible reason for preferring one to the other, for we cannot base a preference either on what we see or on what we know (or believe). Consider, for example, two forest landscapes that look the same or very much the same in terms of features such as species of trees, kinds of ground cover, amount of deadwood, and kinds of wildlife, which landscapes, however, are such that one is sustaining itself while the other is being very slowly destroyed by, for example, acid rain. Without knowledge (or beliefs) about their respective sustainabilities and given that they look alike, we seemingly can have no preferences for one over the other. Thus, we may leave *Case 1*, concluding that it is not interestingly relevant to our focus question.

Now consider *Case 3*. Concerning this case, in which two landscapes look the same and we know which is which, it might be contended that if there are any differences in preferences, then, since the two landscapes look the same, these differences are not differences in *aesthetic* preferences. Given the traditional and narrow understanding of aesthetic preferences for landscapes as preferences for the ways in which landscapes strike our senses, most typically our sense of sight, then any preferences for one or the other of two landscapes that look the same seemingly can not be aesthetic preferences, but rather must be some other kind, for example, moral or economic preferences. For instance, consider again the above mentioned example: Even though the two forest landscapes look alike, we might prefer the sustainable landscape to the non-sustainable one for moral reasons having to do with the duty to maintain wildlife habitat or for economic reasons having to do with the possibility of long term timber yield. Thus, if the preferences involved in such cases are not aesthetic preferences, we would seemingly have grounds for, as with *Case 1*, deeming *Case 3* to also be irrelevant to this discussion. However, as we shall see, unlike *Case 1*, *Case 3* must shortly be reconsidered.

By contrast with *Case 3*, *Case 2*, in which two landscapes look different and we do not know which is which, presents a different kind of problem. Concerning this case it may be argued that although any preferences we might have are indeed *aesthetic* preferences, these preferences are not, since we do not know which landscape is and which is not sustainable, preferences for *sustainable* landscapes, but only for landscapes that have a particular look. Unlike *Case 3*, here we have the right kind of preferences, but not preferences *for* the kinds of landscapes that interests us. In an example similar to that mentioned above, but where the two forest landscapes look different from one another, we might prefer the sustainable landscape to the non-sustainable one, but without knowledge (or beliefs) about its sustainability, our preferences are simply for its appearance, and not, as it were, preferences for its sustainability. Thus, as with *Cases 1* and *3*, we may leave *Case 2*, for it seems to be irrelevant to our focus question of whether or not people aesthetically prefer sustainable landscapes.

However, even if *Case 2* need not be further pursued in this discussion, this kind of case is very important, for it directs attention to another important research question: That of whether or not sustainable landscapes have, *in virtue of their being sustainable*, a particular kind of appearance. For example, is there a unique kind of experience that sustainable landscapes provide or are there clues or signs given in the look of a sustainable landscape by which we can know that it is sustainable? If there is in this sense *an experience of sustainability* or *a look of sustainability*, then in such cases knowing follows from seeing. However, whether or not there is such an experience or look is seemingly an empirical rather than a theoretical question. Thus, I here only highlight this question and do not pursue it. Fortunately, it is pursued by others better equipped to address it (*e.g.* Thayer, 1989; Gobster, 1994; Kimmins, 1999; Kimmins, Chapter 4, this volume).

5   Case 4

This leaves *Case 4*, which therefore appears, at least initially, to be the most interesting kind of case within the scope of this discussion. In this case, in which two landscapes look different and we know which is and which is not sustainable, it may plausibly be thought that people do indeed *aesthetically* prefer *sustainable* landscapes. That is to say that in this case people might well be said to have preferences that are actually aesthetic in nature and are actually for landscapes in virtue of the fact that they are sustainable. However, demonstrating that this is the case requires further investigation - mainly of how our knowledge of (or our beliefs about) a landscape's nature informs our perception of it in such a way that two landscapes that have different natures can strike us as aesthetically different as a function of that knowledge of (or those beliefs about) the differences in their natures (Carlson, 1981; 1995; 2000; Eaton, 1997b). This point needs elaboration.

Such elaboration is best provided by means of examples. First, consider an illustration other than a landscape. For instance, if we know that a building has been poorly constructed with faulty material, then cracks that appear in its foundation will not look aesthetically pleasing to us, they will in fact simply *look bad*. On the other hand, cracks that appear in a piece of antique furniture or on the surface of an old oil painting, cracks that we know to be the result of the natural ageing process of the finish, may strike us as very aesthetically pleasing, as *looking good* and as enhancing the look of the whole object. Now consider landscapes. As with the previous example, we may find the look of the vegetation of one landscape, such as a garden landscape, aesthetically pleasing, if we know that it is the result of careful maintenance and healthy growth.

Yet the different look of the vegetation of another similar landscape may not strike us as so, if we know that it is the result of neglect or blight. Likewise, certain features of forest landscapes, such as dying and decaying trees or trees blackened by fires, may look aesthetically good when we know them to be part of the sustaining process of a sustainable landscape and yet look aesthetically bad when we know them to be indications of the decline and impending demise of a non-sustainable landscape.

The above examples suggest that concerning two landscapes that look different and of which we know (or believe) that one is and the other is not sustainable, we can plausibly be said, in light of our knowledge (or our beliefs), to aesthetically prefer (the look of) the one that is sustainable. However, if this point is granted concerning cases in which two landscapes look different, it can also be extended to apply to cases in which two landscapes look the same. Making this extension requires revising the idea that in such cases, if there are differences in preferences, then, since the two landscapes look the same, these differences can not be differences in *aesthetic* preferences. The first step in revising this idea is noting that the previously introduced traditional and narrow interpretation of aesthetic preferences as preferences only for the ways in which things strike our senses is too narrow. There is an equally traditional and slightly less narrow understanding of aesthetic preferences in which they are taken to be preferences not simply for the look of things, but also for what such things *express* to us in light of our knowledge (or our beliefs) about why they look as they do (*e.g.* Sagoff, 1974; Carlson, 1976). On this less narrow understanding of aesthetic preferences, genuine aesthetic preferences can be a function not only of the look of things but also of what they express in light of what we know (or believe) about them.

Given this less narrow understanding of aesthetic preferences that encompasses not only the look of things but also what they express, the illustrations introduced above may be described in a slightly fuller way. Consider the example in which cracks in a building's foundation do not look aesthetically pleasing to us while those in an oil painting do. It can now be characterized as follows: On the one hand, the cracks in the foundation look bad to us in part because they are expressive of faulty construction and materials and, provided that we know about the faultiness of the construction and materials, the cracks express this to us. On the other hand, the cracks in the old oil painting look good to us in part because they are expressive of a natural ageing process, a process that we know to be a means by which such paintings achieve their distinctive look. The upshot is that we aesthetically prefer the cracked look of the painting, but not that of the foundation, and our aesthetics preferences are a function of what is expressed in light of both our seeing and our knowing.

Given these observations, consider again the illustrations of landscapes. In the example in which we find the look of the vegetation of one landscape, which we know to be carefully maintained, aesthetically pleasing and the different look of the vegetation of another similar but neglected landscape not so, we may conclude that this is in part because the vegetation of the first landscape expresses care and concern, while that of the second expresses neglect. Likewise, consider the example in which certain features of forest landscapes, such as dying and decaying trees, look aesthetically good when we know them to be part of the sustaining process of a sustainable landscape and yet look aesthetically bad when we know them to be indications of the demise of a non-sustainable landscape, such as, in the example of a forest being slowly destroyed by acid rain. Here we can characterize the first landscape simply as expressing its sustainability and the second simply as expressing its non-sustainability. We can thus conclude that, when we aesthetically prefer sustainable landscapes, we aesthetically prefer them because they are sustainable *and* because they *express their sustainability*. In short, part of what is involved in our aesthetic preferences for sustainable landscapes is the fact that we know (or believe) them to be sustainable and that, in light of this knowledge (or these beliefs), they express their sustainability to us.

6   Case 3 Reconsidered

In light of this conclusion, we can now return to and directly reconsider *Case 3*, which we initially discussed and then set aside in Section 4. Recall that this is the case in which two landscapes look the same and we know which is and which is not sustainable. Concerning this case we tentatively suggested in Section 4 that if there are any differences in preferences, then, since the two landscapes look the same, these differences are not differences in *aesthetic* preferences. However, in light of the observations in Section 5 about a wider understanding of aesthetic preferences and about expression, this tentative suggestion must now be revised. This is because the upshot of these observations is that, given that we know (or have the relevant beliefs) about the differences between two different landscapes, such landscapes may express different things to us *even if they look exactly alike*. How is this possible?

To appreciate the answer to this question, consider once again some illustrations, starting with an example other than a landscape. Take the example of art. Concerning works of art it is generally accepted that even if two works are identical in their physical appearances and in this sense look exactly the same, they can yet express very different things, if they have, for example, different ages, natures, and/or means of production (*e.g.* Walton, 1970). Consider, for example, a pair of apparent works of art, which although identical in appearance, are such that one is an authentic hand-painted work of an old master while the other is a recent machine-crafted product of a modern forger. Clearly these two objects express different things, and if we know which is which and thus what each expresses, we will most likely aesthetically prefer the former to the latter, even though they look exactly the same to us (Sagoff, 1976). The same is true of natural and artificial landscapes (Carlson, 1981; 2000). Consider, for instance, the example of ecological restoration. It seems clear that even though a well executed ecologically restored landscape and a original natural landscape may look exactly (or almost exactly) the same, they will express different things, and thus, with knowledge of their differences, there will be grounds for aesthetically preferring one to the other (Elliot, 1982; 1997).

Now consider yet again the previously described illustrations of sustainable and non-sustainable landscapes. First, think about the example in which we find the look of the vegetation of a carefully maintained landscape aesthetically pleasing and the look of the vegetation of another similar but neglected landscape not so, because the vegetation of the first landscape expresses care and concern, while that of the second expresses neglect. Now note that this could be the case *even if the vegetation of the two landscapes looked just the same to us*, provided we know which is carefully maintained and which is neglected. Likewise, consider the example of forest landscapes in which dying and decaying trees look aesthetically good when we know them to be part of the sustaining process of a sustainable landscape and yet look aesthetically bad when we know them to be indications of the demise of a non-sustainable landscape. In this example, even if the dying and decaying trees look exactly the same in the two different landscapes, in the first landscape they nonetheless express the sustainability of that landscape while in the second they express non-sustainability. Thus, even though the two landscapes look identical to one another, we have grounds for aesthetically preferring the former to the latter.

In light of these examples, we may conclude not only, as noted above, that when we aesthetically prefer sustainable landscapes, we aesthetically prefer them because they are sustainable *and* because they *express their sustainability*, but also that we can plausibly be said to aesthetically prefer landscapes that express their sustainability even if they look no different from landscapes that do not. Thus, in response to our focus question, Do people prefer sustainable landscapes?, we may say the following: Although whether or not people in general do in fact prefer sustainable landscapes is an empirical matter, it can yet be concluded that they may plausibly be said to aesthetically prefer such landscapes, *provided, of course, that they know (or believe) that they are sustainable*. In short, our aesthetic preferences for sustainable landscapes are a function of both our seeing and our knowing.

7 Conclusion: Two Ramifications

I conclude this discussion by noting two ramifications of this kind of response to the focus question. The first is that, since our aesthetic preferences for sustainable landscapes are grounded in large part in what they express, in light of our knowledge of (or our beliefs about) their sustainability, there is a natural alignment of our aesthetic preferences with what may be called our ethical preferences. This is because in such cases those landscapes we aesthetically prefer will typically be ones that express things we ethically prefer (Carlson 1976; 1981). For example, when we aesthetically prefer a sustainable landscape, such as, for instance, a carefully maintained landscape garden, because it expresses sustainability, this is in some measure because such a landscape also expresses things such as care, concern, and prudence. In short, sustainable landscapes typically express a number of virtues, which are involved with sustainability and for which we rightly have an ethical preference. By contrast non-sustainable landscapes express not only their non-sustainability, but frequently also express vices such as waste, greed, exploitation - things that we rightly ethically despise. The upshot is a deep and significant linkage between the aesthetics and the ethics of our relationship with our landscapes - a linkage between environmental aesthetics and environmental ethics (Eaton, 1997a; Saito, 1998; Carlson, 1998).

The second ramification follows directly from the fact that people can aesthetically prefer sustainable landscapes even if such landscapes look exactly the same as non-sustainable landscapes. The ramification is that people can aesthetically prefer sustainable landscapes to non-sustainable ones even if the former look (at least in the short run) as aesthetically *bad* (in the narrow sense of aesthetic) as the latter. For this to be the case, what is required is that we know which landscapes are sustainable and which are not and come to appreciate what such landscapes express in light of this knowledge. In short, what is required is that we not only see a landscape, but also know its true nature. It should be obvious that this ramification has relevance to what might be called applied aesthetics of landscapes (Carlson, 2000) and, therefore, to the work of those who plan, manipulate, and thus determine our present and future landscapes. For example, it may have application to practices such as what is called New Forestry, which is sometimes accused of procedures, which, although claimed to be sustainable, are seen as having aesthetically bad short term results.

This second ramification underscores the fact that our knowledge of the sustainability of a landscape can be more important to our aesthetic preferences than how it happens to look at any particular point in time. Of course, this gap between what we see and what we know might be bridged by a positive answer to the important question highlighted in Section 4: The question of whether or not sustainable landscapes have, in virtue of their being sustainable, a particular kind of look. As noted, if there is in this sense an identifiable *look of sustainability*, then seeing and knowing converge in that the latter can follow from the former. However, as long as any split remains between what we see and what we know, we must keep in mind that our knowledge of the sustainability of a landscape is frequently more significant to our aesthetic preferences than how it happens to look. In short, concerning landscapes, as with most things, in the last analysis reality is more important than appearances.

References

Bell, S. (1999) *Landscape: Pattern, Perception and Process*. Routledge, London.

Bourassa, S.C. (1991) *The Aesthetics of Landscape*. Belhaven, London.

Brundtland, G.H. (1987) *Our Common Future*. Oxford University Press, New York.

Carlson, A. (1976) Environmental aesthetics and the dilemma of aesthetic education. *Journal of Aesthetic Education* 10: 69-82.

Carlson, A. (1977) On the possibility of quantifying scenic beauty. *Landscape Planning* 4: 131-72.

Carlson, A. (1981) Nature, aesthetic judgment, and objectivity. *Journal of Aesthetics and Art Criticism* 40: 15-27.

Carlson, A. (1990) Whose vision? Whose meanings? Whose values? Pluralism and objectivity in landscape analysis. In: P. Groth (ed) *Vision, Culture, and Landscape: The Berkeley Symposium on Cultural Landscape Interpretation.* Berkeley: Department of Landscape Architecture, University of California.

Carlson, A. (1993) On the theoretical vacuum in landscape assessment. *Landscape Journal* 12: 51-6.

Carlson, A. (1995) Nature, aesthetic appreciation, and knowledge. *Journal of Aesthetics and Art Criticism* 53: 394-400.

Carlson, A. (1998) Aesthetic appreciation and the natural environment. In: S. Armstrong and R. Botzler (eds) *Environmental Ethics: Divergence and Convergence,* Second Edition. McGraw Hill, New York.

Carlson, A. (2000) *Aesthetics and the Environment: The Appreciation of Nature, Art and Architecture.* Routledge, London.

Eaton, M.M. (1997a) The beauty that requires health. In: J.I. Nassauer (ed) *Placing Nature: Culture and Landscape Ecology.* Island Press, Washington, DC.

Eaton, M.M. (1997b) The role of aesthetics in designing sustainable landscapes. In: Y. Sepanmaa (ed) *Real World Design: The Foundations and Practice of Environmental Aesthetics.* University of Helsinki, Helsinki.

Elliot, R. (1982) Faking nature. *Inquiry* 25: 81-93.

Elliot, R. (1997) *Faking Nature: The Ethics of Environmental Restoration.* Routledge, London.

Elsner, G.H. and Smardon, R.C. (eds) (1979) *The Proceedings of Our National Landscape: A Conference on Applied Techniques for Analysis and Management of the Visual Resource.* USDA Forest Service Pacific Southwest Forest and Range Experimental Station, Berkeley, CA.

Gobster, P.H. (1994) The aesthetic experience of sustainable forest ecosystems. In: W.W. Covington and L.F. DeBano (eds) *Sustainable Ecological Systems: Implementing an Ecological Approach to Land Management.* USDA Forest Service Rocky Mountain Forest and Range Experiment Station, Fort Collins, CO.

Kaplan, R. and Kaplan, S. (1989) *The Experience of Nature: A Psychological Perspective.* Cambridge University Press, Cambridge.

Kimmins, J.P.(Hamish) (1999) Biodiversity, beauty, and the "beast": are beautiful forests sustainable, are sustainable forest beautiful, and is "small" always ecologically desirable? *Journal of Forestry* 75: 955-60.

Nassauer, J.I. (ed) (1997) *Placing Nature: Culture and Landscape Ecology.* Island Press, Washington.

Sagoff, M. (1974) On preserving the natural environment. *Yale Law Journal* 84: 205-67.

Sagoff, M. (1976) The aesthetic status of forgeries. *Journal of Aesthetics and Art Criticism* 35: 169-80.

Saito, Y. (1998) The aesthetics of unscenic nature. In: A. Berleant and A. Carlson (eds) *Special Issue: Environmental Aesthetics. Journal of Aesthetics and Art Criticism* 56: 101-11.

Thayer, R.L. (1989) The experience of sustainable landscapes. *Landscape Journal* 8: 101-10.

Walton, K. (1970) Categories of art. *Philosophical Review* 79: 334-67.

Zube, E.H., J.L. Sell, and J.G. Taylor (1982) Landscape perception: research, application and theory. *Landscape Planning* 9: 1-33.

Chapter four:

Visible and Non-Visible Indicators of Forest Sustainability: Beauty, Beholders and Belief Systems

J.P. (Hamish) Kimmins
Department of Forest Sciences, University of British Columbia
Vancouver, British Columbia

1   Introduction

Humans generally live a short time relative to the time-scales of forest ecosystems. As a result, we are frequently unaware of the change that is an inevitable feature of forests. Many people do not take kindly to change, especially environmental change: the Peter Pan Principle in resource conflicts (Kimmins, 1997a). Combined with their limited experience of natural change in forests, this can lead many people to the philosophy that change is bad, whether it is caused by forest management or natural processes.

Like other animals, we assess our environment largely through our senses, of which for most people the visual sense is the most powerful (Kimmins, 1999). Both beauty and the evaluation of ecosystem condition and sustainability are literally in the eye of the human beholder when it comes to forests. Much of the change that occurs in forests, whether this is the result of human actions or natural (non-human) ecological events, is initially not pleasing to the aesthetic senses of many people. This further supports the negative impressions of change in forest ecosystems.

Our inherent dislike of visual change, combined with various belief systems about nature (Kimmins, 1993), leads many to judge the ecological condition and sustainability of landscapes in a way that is sometimes inconsistent with the maintenance of a variety of desired forest values. "Snapshot" visual assessments of the present condition of a forest landscape are sometimes an accurate predictor of the supply of future values from that landscape, but in many cases this is not true. Neat and tidy forestry may be judged visually to be good stewardship, but can sometimes deplete soil fertility and reduce wildlife habitat and species diversity. Minimal disturbance, *soft touch* harvesting may look like good stewardship to many people, but can reduce functional and species diversity, and eliminate temporal diversity in some kinds of forest ecosystems, even though it may increase some aspects of structural diversity.

Evaluation of the visual appearance of ecosystems must be separated from the assessment of concepts such as sustainability, biodiversity and ecosystem *health* and *integrity* if we are to be successful in selecting forest management methods that will achieve our management objectives. Current visual resource values are important, but should be balanced against longer-term aesthetics and our ethical obligations to pass on a diversity of values to future generations. We must balance the desire to satisfy our current aesthetic sensibilities against our desire to create a diverse legacy of biological and social values.

In this chapter I will explore the key components of ecosystem sustainability and evaluate the degree to which visual clues can provide an accurate assessment of sustainability and related concepts. Beauty will always be in the eye of the beholder, but we must recall the childhood stories of *Beauty and the Beast* and the *Ugly Duckling*. In nature, beauty sometimes comes out of ugliness, and sometimes ugly stages of forest development may be a necessary pre-requisite for later beauty. We must not allow belief systems about nature that assert that if something looks ugly it must inevitably be ecologically bad to become the basis for forest management and conservation strategies. Aldo Leopold, in a much-quoted passage from *The Land Ethic* (1953), drew attention to the possible relationship between beauty and ecosystem condition; beauty and stewardship.

> A thing is right when it tends to preserve the integrity, stability and beauty of the biotic community. It is wrong when it tends otherwise.

However, on the next page he advanced the following, rarely quoted, warning:

> Conservation is paved with good intentions which prove to be futile, or even dangerous, because they are devoid of critical understanding either of the land or of economic land use.

By this second quote he made it clear that his use of beauty referred as much to ecological beauty as to any aesthetic sense of beauty, and that stability is a dynamic term involving non-declining patterns of change.

2   Some Basic Concepts and Definitions

2.1 Ecosystems

A terrestrial ecosystem is any ecological system that exhibits the attributes of:

Structure - plants (or some other primary producer), animals, microbes, soil, and atmosphere (medium)

Function - energy flow, nutrient cycling and regulation of hydrological cycles

Complexity - the complexity of events in complex systems renders them difficult to predict, unless one has a good knowledge of the components and processes, and their interactions.

Interaction of its components - it is a system in which there is a high degree of linkage between the components through a variety of processes.

Change - ecosystems are continually changing, and exhibit great variation in structure, function, complexity and the interaction of their components from time to time and from place to place (Kimmins, 1997b).

A forest ecosystem is an ecosystem dominated by trees. It is more than just the trees, however; it is the trees, other plants, forest microbes and animals, the forest floor, the forest soil and the forest microclimate. We must not allow a focus on the trees to prevent us from seeing the forest ecosystem. It is still a forest if you remove the trees and temporarily lose the forest microclimate, because it still has the other forest organisms and the forest soil. However, it only remains a forest if the processes of change that restore tree cover operate over a short enough time-scale to re-establish the dominant influence of trees on soil, microclimate, hydrological cycles, biogeochemical cycles and ecosystem productivity before the forest soil and forest minor vegetation characteristics are lost and replaced by non-forest soil and minor vegetation characteristics. If these processes are sufficiently delayed, the forest ecosystem may become a meadow ecosystem or a shrub ecosystem until forest processes once again control the site and return it to a forest ecosystem.

2.2 Biodiversity

The structure and function of forest ecosystems vary from place to place. We call this biological diversity. There are many measures of biological diversity: genetic (within a species), taxonomic (diversity of taxonomic groups), species (richness, evenness), structural, and functional. The diversity of measures of biodiversity complicate any discussion of the topic, and make it difficult to identify both what we mean by, and what we should do to conserve biodiversity.

Biological diversity at the local and regional landscape scales is a direct reflection of ecological diversity - the diversity of *ecological stages* (physical environments defined by climate, soils and topography) on which the *play of nature* (change in ecosystems over time due to disturbance and other ecological processes) is acted out by the *ecological actors* (the species). The particular ecological play one can observe at a particular location in a landscape is determined by the ecological stage on which it is occurring at that location.

The list of ecological actors that are present on the ecological stage is determined not only by the ecological play that is being acted out, but also by which Act of the play is currently on stage. The mix of ecological actors and the ecological story changes over the course of the ecological play. This reflects that most important measure of biological diversity noted above - temporal diversity. All other measures of biological diversity change over time, especially at smaller spatial (alpha) scales. Biodiversity is not a fixed, equilibrium concept, and, consequently, it is difficult to know what is meant by such rallying cries such as "save biodiversity", "protect biodiversity", and "preserve biodiversity", and what should be done to act on such admirable and popular directives.

Shakespeare said, "all the world's a stage", even though his plays were generally performed on very small stages. Similarly, the theatre of nature can be considered at various different spatial scales from very local to very large scale, and so can the measures of biological diversity.

All the above-mentioned measures of forest biodiversity can be evaluated in a local stand (*e.g.* a 1 - 100 ha forest area), a local landscape (*e.g.* 100 - 10,000 ha of forest) or in a geographical region or large-scale landscape (*e.g.* >10,000 ha). Biodiversity at these different scales is referred to as alpha, beta and gamma diversity, respectively. Measures of alpha diversity typically change significantly over time (high temporal diversity). Temporal diversity tends to be progressively less at local (beta) and regional (gamma) landscape scales but, even at the large scale, there may be significant temporal diversity in types of forest that are subject to large scale disturbance such as fire or insect epidemics.

The relative importance of the different individual measures of biodiversity in terms of overall biodiversity conservation varies at different spatial scales, as do the temporal patterns of change in these measures, from local stands, to local landscapes, to regional landscapes. The interaction of the spatial and temporal aspects of biological diversity with the many measures thereof makes it difficult to relate instantaneous (snapshot) visible indicators of biodiversity to what the future time trends and the sustainability of these measures will be.

2.3 Ecological Function and Ecological Change

Ecosystems are energy processing and storage systems. They function by capturing sunlight energy and storing it as the valency energy of complex organic molecules. To do so they need the chemical building blocks of these molecules; the nutrient elements. Ecosystem function is determined by leaf area and the efficiency with which the leaves catch, convert and store the sunlight energy. Both of these are determined by the availability of moisture and nutrients and the ability of plants to acquire them. The functional role of plants in capturing and storing energy and cycling nutrients is divided up between different plant species according to who gets there first, and who is able to survive, compete for resources and grow under the prevailing conditions of microclimate and resource availability. It also depends on the life history details of size, longevity, competitiveness, rate of growth, resource needs and susceptibility to agents of mortality. Over time, the functional tasks (the ecological roles) of the ecosystem are performed (acted out) by different species, which successively occupy and are then replaced on the ecological stage. As this changing of the ecological guard occurs (ecological succession), there are changes in ecosystem structure, function and complexity (sometimes increases, sometimes decreases), with concomitant changes in measures of biological diversity. However, the integrity of the forest as an ecological system is normally sustained through this period of naturally-driven change, despite the periodic loss of the integrity of any particular group of species by natural processes of replacement (Kimmins, 1996).

In most of nature's ecological plays, there is a cast of thousands. Forest ecosystems are generally characterized by many different species. For plants and vertebrate animals, this is especially true in humid tropical forests, and the diversity of these types of organism tends to decline as one goes towards the poles or up in elevation. For microbes and invertebrates, latitudinal and elevational trends are less well documented or unknown. Where there are many species in an ecosystem, the *ecological tasks* (ecological roles) are divided between the several to many species in each functional ecological group (guilds), and many or even most of the species in a guild may be able to complete their basic functional ecosystem *task* without assistance from all the other guild species that could exist there. There is thus considerable ecological redundancy in many forest ecosystems - many species may be missing from a guild with little perceptible effect on overall ecosystem function in spite of changes in measures of biodiversity such as species richness. In other forest ecosystems (*e.g.* many humid tropical forests) there may be such intimate interdependencies between species (*e.g.* particular plant species and particular pollinating insects) that loss of one species may result in the eventual loss of one or more other species.

Some northern ecosystems with relatively low species diversity can be as productive as, or more productive than, some highly diverse tropical systems. However, ecosystems with many species in each ecological group (guild) will function differently from those with few species; they may be either more or less productive, depending on how you define productivity, and they may be more or less stable and resilient, depending on the definition of these terms. The wide range in measures of biological diversity in both productive and unproductive ecosystems, and the ability of a particular ecosystem to function in spite of a wide variation in species and structure, renders it very difficult to make useful generalizations about the relationship between measures of diversity and concepts such as sustainability, stability, resilience, and ecosystem health and integrity (Kimmins, 1997c).

Generally, there is a characteristic historical range in measures of diversity in any particular forest ecosystem type over time, and it may be necessary for the ecosystem to fluctuate over this full historical range as time passes since the last ecosystem disturbance if it is to function as it has historically.

Periodically the curtain comes down on the ecological play. Fire, wind, insects, disease or landslides interrupt ecosystem development, changing the species and the ecosystem condition, clearing the ecological stage, and making way for the start of a new ecological act, a new ecological play, or a repeat of the previous ecological performance. Where these disturbance events result in a different set of species occupying the site, the ecological play may change, but it will remain within the limits set by the ecological stage (Attiwill, 1994).

3   Stand-Level Sustainability as a Non-Declining Pattern of Change: The Concept of Ecological Rotations

The characteristic of change over time is as fundamental to the *health* and *integrity* of stand-level ecosystems as are the characteristics of structure, function, complexity and interaction of the parts. Change over time in the character of any particular ecosystem as a result of disturbance and ecosystem recovery is as inevitable as the fact that ecosystems are different from place to place as a result of ecological diversity (Botkin, 1990; Attiwill, 1994). However, the characteristic of temporal change poses a dilemma for the stand-level assessment of sustainability. If ecosystems are continually changing in structure, function and diversity rather than being constant, (as in an equilibrium system), then what does ecosystem sustainability mean?

It would seem that the only logical way by which to define sustainability at the stand or local ecosystem spatial scale is to consider it as *a non-declining (or non-increasing) pattern of change*. A change in ecosystem attributes is not a threat to sustainability as long as the combination of degree of change, the frequency of change and the rate of recovery of the ecosystem back to its original or to some desired new condition causes it to fluctuate round some long-term mean value. This approach to the definition of sustainability is referred to as the concept of *ecological rotations*: the time required for a specific ecosystem value or condition to recover to its pre-disturbance level, or to some desired new condition, following a particular type, extent and severity of disturbance (Kimmins, 1974).

Based on the concept of ecological rotations, there is no fixed formula for the assessment of sustainability. If frequent change is to be sustainable, it must be relatively small-scale and/or low severity change, unless the ecosystem recovers very quickly. For example, with partial (*e.g.* selection) harvesting in which you enter the forest to remove trees perhaps every ten years, great care must be taken to avoid soil damage, such as compaction, and damage to regeneration and the roots of live trees. If a high degree of ecosystem change is to be sustainable, it should generally be infrequent, but the acceptable frequency will depend on the resilience of the ecosystem or ecosystem value in question. For example, clearcutting and burning which can substantially alter several ecosystem characteristics will normally require re-entry periods of 80 - 120 years in many British Columbia forests, although on productive and resilient sites a rotation length of 60 - 80 years may be sustainable of at least some ecosystem values.

Evaluation of ecosystem condition within a sustainability paradigm should be based upon the concept of ecological rotations and be monitored on the basis of desired "temporal fingerprints of change" (Kimmins, 1990). A temporal fingerprint is the desired or anticipated temporal pattern of change in ecosystem attributes based on some selected combination of disturbance type, scale, severity and frequency for an ecosystem of particular resilience. The actual degree of ecosystem disturbance should be related to the ecology of desired values. For example, deciduous hardwoods in British Columbia are generally disturbance-related, shade intolerant pioneers that have great value for biodiversity, aesthetics and site fertility. They require a higher level of ecosystem disturbance than is sustained by *soft touch* or *ecoforestry* that looks nice; a level of disturbance that will generally require a low frequency of disturbance.

4   Landscape-Level Sustainability as a Shifting Mosaic of Ecosystem Conditions

In contrast to stand-level sustainability, landscape-level sustainability in many forest regions refers to an *overall* regional mosaic of forest conditions (different forest ages and different stages of forest development) that may remain relatively constant (or fluctuate between historical limits of variation), but in which the physical location of any particular forest age or condition changes over time: a shifting mosaic of reasonably constant overall character (Kimmins and Duffy, 1991). In some forest regions, the overall composition of the landscape mosaic may remain constant over spatial scales of thousands to tens of thousands of hectares. In landscapes subject to large fires (*e.g.* parts of the boreal forest), a sustainable landscape mosaic may require several million hectares (Andison and Kimmins, 1999; Bunnell and Huggard, 1999).

Because landscape sustainability is a shifting mosaic often covering very large areas, it may be very difficult to visualize. Most people will only observe a small proportion of the landscape at any one time, giving the impression of non-sustainability. This has been the case in forest harvesting in British Columbia. Considered within a landscape of a million hectares being harvested at a sustainable rate, a clearcut of several thousand hectares may be part of a shifting sustainable mosaic that may mimic the historical disturbance scale of large wildfires. Seen as part of a 5,000 ha valley landscape, the clearcut would appear to be non-sustainable.

5 Visible and Non-Visible Indicators of Sustainability

5.1 Visible Indicators

Nothing is simple in life, and even the distinction between visible and non-visible requires elaboration. Visible to whom: the uninterested, an urban dweller, an environmentalist, a forest worker, a technical expert or a forest scientist? Visible at what spatial scale and from what distance? Visible from a valley bottom or a mountain top; from a car window or a helicopter window, from a boat or from an airplane? Visible with the naked eye, with binoculars or microscope, or with an increment borer, a spade or an axe? Visible in still pictures, in video, or only in time-lapse photography spanning many years or decades? Visible in a telephoto, wide-angle or a fisheye photograph? Different visual information is available to different people with different knowledge, experience, belief systems and paradigms. Thus, what is visible to one person may be invisible to another. What is visible by one technology or methodology may not be visible by another.

The diversity of factors that determines visibility makes it difficult to assign indicators of ecosystem condition and sustainability to either visible or non-visible categories. For example, a forest ecologist, biologist or soil scientist, or an experienced field naturalist or outdoorsperson, may be able to make informed judgements about sustainability by a visual inspection of the plants and soil with the aid of a shovel, an axe (for investigation of bark beetles and root/stem diseases) and an increment borer. To the average urbanite, the visual indicators so revealed may well be *invisible*; they would be uninterpretable and not recognized as indicators (see Plates 2 and 3).

In the following discussion, I will assume that visual indicators refer to what the average person (whoever this might be) could be expected to observe from a casual inspection of a forest stand from the roadside, and of a landscape from a ridge top. I will refer to these as *easily visible features*.

There are very few, if any, easily visible biophysical indicators that, on their own, can be used to judge sustainability at the stand level. Without reference to the temporal fingerprint of sustainable, non-declining change for that particular ecosystem, it is not possible to put the existing easily visible features of an ecosystem into the context of sustainability. Without knowledge of (1) the frequency of future disturbances to that ecosystem, and (2) the rate of ecological processes (most of which cannot easily be seen) that will take the ecosystem back to, or towards, its pre-disturbance or long-term average condition, it is not possible to evaluate the consequences for stand level sustainability of a given, instantaneous visual condition of the ecosystem.

Easily visible features of a forest stand can indicate whether that forest conforms to some simple definition of the desired present structure, species composition or condition for that ecosystem. By themselves, these features on their own cannot tell you whether that ecosystem is being managed sustainably, however. Certainly, a simple visual inspection of the forest may tell you whether or not the forest in its present condition will support certain current values or not; and a detailed evaluation of present species composition, stand structure, regeneration, the forest floor, the mineral soil and the inventory of standing dead trees and decaying logs can be interpreted in terms of future ecosystem conditions. However, without a knowledge of the expected future disturbance regime and the rates of processes that cannot readily be observed (but may be interpretable by a knowledgeable and experienced individual from easily observed features), it is difficult to predict stand-level sustainability with confidence. Thus, visual indicators are only useful at the stand spatial scale if used in conjunction with the concepts of ecological rotation and temporal fingerprints of change.

Clearly, the more a person understands about ecosystem structure and function, the ecological role of disturbance, and the ecosystem characteristics and processes that determine its recovery from disturbance, the more a person will be able to observe, understand and interpret the indicators of ecosystem condition and future development. Without this knowledge an observer may see ugliness and destruction; with this knowledge they may see an ecosystem passing through one of its stages of change within the range of variability that is consistent with sustainability.

In contrast to the difficulties with visual assessment at the stand level, it is easier to assess sustainability at the landscape level, but only if you can see the landscape at the appropriate spatial scale. Given that one can see the appropriate landscape, one can evaluate whether or not the range of age classes and forest conditions conform to a sustainable shifting mosaic, and if the landscape pattern of the mosaic is consistent with the values desired from that landscape (Kimmins and Duffy, 1991). An inventory of the present age classes and timber volumes in the forest provides part of the basis for predicting landscape sustainability of timber-related values, but is not sufficient on its own. An evaluation of sustainability also needs to know the *operability* of timber values as affected by their spatial patterns, and the inventory and spatial arrangement of non-timber values. However, even this evaluation is incomplete. To know if the landscape is truly being managed sustainably, one also needs to know if each stand in the landscape is being managed sustainably. Thus, visual evaluations can provide a significant component of sustainability analysis at the landscape scale, but not a total assessment which requires both landscape and stand level evaluation.

Much of the recent debate over forest sustainability has focused on vertebrate wildlife; spotted owls, salamanders, deer, moose, caribou, cavity-nesters, etc. (Kimmins, 2000). The ecosystem management paradigm is strongly based on retaining snags, wildlife trees, coarse woody debris and connectivity corridors of mature vegetation for the habitat needs of these vertebrates. Much less attention has been focused on non-vertebrates. Because of the generally inadequate knowledge of the ecology of even the major vertebrate species, the design of forestry to sustain wildlife has, of necessity, been based more on opinion than on empirical data. This can, unfortunately, lead to the assumption that what looks good to us will also be good for the wildlife, which may or may not be true. There is evidence that some *wildlife corridors* may become *predator traps*, where the funnelling of the prey wildlife species may increase their exposure to predation. Scattered wildlife trees in harvested areas may look like suitable bird habitat, but may lack thermal protection and protection from predators. One wonders if lack of adequate data on habitat suitability for most wildlife species results in decisions concerning conservation of their habitat being based as much on human-based aesthetic judgements as on good ecology.

5.2 Non-Visible Indicators

As noted above, easily visible indicators are only useful in the context of the concepts of ecological rotations and temporal fingerprints of change. These two concepts involve several non-visible indicators of sustainability.

A key component of both concepts is the frequency and severity of ecosystem disturbance. The risk of wind, fire and insect disturbance, and the intended timber harvest rotation length are not easily visible indicators, yet are key determinants of sustainability. Decisions about rotation length have in the past been mainly based on market and economic criteria, which again are not visible and yet play a major role in determining stand level sustainability. They also change over time.

A critical component of sustainability is the inventory of organic matter and nutrients in the forest. To the experienced, knowledgeable and informed observer, there are visual indicators in the mineral soil, organic forest floor and coarse woody debris from which one can infer soil organic matter and nutrient status, but even these individuals may require chemical analysis to confirm the inferences from the visual indicators. To the uninformed and inexperienced observer, the soil holds many secrets that cannot easily be revealed.

Sustaining long-term site productivity in most northern and temperate forests depends on maintenance of ecosystem nitrogen levels and dynamics. Forest management that maintains late seral conditions which are visually attractive may reduce nitrogen inputs from nitrogen fixing species such as red alder, or from species such as birch or aspen that make non-symbiotic contributions of nitrogen to the ecosystem, and can lead to stagnation of nitrogen cycling and reduced ecosystem productivity. Early seral hardwood species are important for wildlife habitat and measures of biodiversity as well as for soil fertility. Ensuring that these species are retained within the landscape may require severities, scales and frequencies of disturbance that are aesthetically displeasing to some observers.

6   Tools for the Assessment of Forest Sustainability

The complexity of sustainability issues in forestry requires the use of a variety of analytical and decision-support tools. The first of these should be an ecologically-based site classification system. Another is ecosystem management models.

One of the difficulties in answering the question "What is sustainable forestry?"

is that the answer will vary in different types of forest (Kimmins, 2000). This reflects the ecological diversity of forests, as well as their biological diversity. Before one can address this question, it is necessary to stratify (classify) the forest landscape by climate, soil, geology, topography and vegetation. This ecological forest classification will identify relatively homogeneous ecological landscape units for which the question can be addressed.

With such a classification in place, we can then develop, calibrate and use ecosystem management models at the stand and landscape scales (Seely *et al.*, 1999; Kimmins *et al.*, 1999a). Such models can evaluate ecological rotations, and project ecological *fingerprints of change*. These forecasting tools cannot, of course, predict the future with accuracy. There are many processes, natural events, and social changes that may occur over the next 200 years which we cannot predict accurately. Also, our knowledge of ecosystems remains incomplete, and our ability to represent current knowledge in models remains inadequate. Consequently, a forecast produced by an ecosystem management model represents one of many possible futures. However, such forecasts do represent the logical consequences of our current knowledge and assumptions as represented in the model, and they can allow us to rank the possible outcomes of different management strategies, if not the absolute values of future conditions. One finds from such modelling that the rankings depend on which particular social or environmental values the ranking is based on; they are usually different for different values.

A major shortcoming of ecosystem management models is that their output is generally graphical and tabular. This format does not communicate effectively to the general public the vision of the future forest foreseen by the models. To use such models in conjunction with public input processes, we must convert the output of simulation models into as visual a form as possible. In spite of the difficulties in identifying visual criteria of sustainability as discussed above, we must nevertheless struggle to put our forecasts of management outcomes into visual format. After all, this is inevitably how the public will judge forestry.

The challenge, then, is for foresters, ecologists and modellers (these are not mutually exclusive professions) to work with experts in visualization to represent our best current understanding of sustainability issues in a visual format, and to communicate to the public in other ways those aspects of sustainability that do not lend themselves to simple visual evaluation. Another challenge, through education, is to make non-visible indicators more visible by increasing the public's understanding of the ecology of forests so that they can balance their emotional judgement of messy ecosystems with an ecological judgement (Nassauer, 1995; Peterken, 1996).

7 Conclusions: How Can We Assess Sustainability?

The public has the right, and the duty, to be concerned about the quality of forest management. The evolution of *green* certification provides a mechanism by which the public can be given a level of assurance about how forests are being managed. Certification in turn requires criteria and indicators. The Forest Stewardship Council (Kimmins, 1997d) has made it clear that their certification does not specifically focus on sustainability; it addresses *good stewardship*, which in turn is defined by ten global principles, and a considerable number of criteria and indicators that are subject to modification by local definitions. In contrast, the Canadian Standards Association certification system focuses on sustainable forest management (SFM) and continuous improvement towards this goal where the present forest, managed *or* unmanaged, does not conform to currently accepted principles, criteria and indicators of SFM. Certification of both *good stewardship* and SFM requires the ability to forecast the consequences, for many different social and environmental values, of the management choices we make at both the stand and landscape scales.

Because some certification schemes aspire to certifying sustainability, and because visual evaluations are quite inadequate for stand level assessments and incomplete for landscape level assessments, mechanisms - planning tools - are needed that are capable of providing forests that are the basis for defensible statements about sustainability.

The first essential planning tool is ecological site classification. This stratifies the ecological diversity of the landscape (climates, soils, topography) into reasonably homogeneous units. These are the *ecological stages* referred to above. They can be recognized visually by plant species composition and soil features. The second tool is ecosystem management models. These should include stand-level ecosystem models such as FORCEE (see Plate 1), FORECAST (Kimmins *et al.*, 1999b), and landscape models in which the temporal development of the individual stands that make up the landscape are defined by stand level ecosystem models. Such a model is under development: HORIZON. It is hard to imagine meaningful certification in the absence of such tools.

Acknowledgements

I would like to thank Maxine Horner for her typing and endless assistance.

References

Andison, D.W. and J.P. Kimmins (1999) Scaling up to understand British Columbia's boreal mixedwoods. *Environmental Reviews* 7: 19-30.
Attiwill, P.M. (1994) The disturbance of forest ecosystems, the ecological basis for conservation management. *Forest Ecology and Management* 63: 247-300.
Botkin, D.B. (1990) *Discordant Harmonies. A New Ecology for the 21st Century*. Oxford University Press, New York.
Bunnell, F.L. and D.J. Huggard (1999) Biodiversity across spatial and temporal scales: problems and opportunities. *Forest Ecology and Management* 115:

113-126.

Kimmins, J.P. (Hamish) (1974) Sustained yield, timber mining, and the concept of ecological rotation: a British Columbian view. *Forestry Chronicle* 50: 27-31.

Kimmins, J.P. (Hamish) (1990) Monitoring the condition of the Canadian forest environment: the relevance of the concept of "ecological indicators". *Environmental Monitoring and Assessment* 15: 231-240.

Kimmins, J.P. (Hamish) (1993) Ecology, environmentalism and green religion. *Forestry Chronicle* 69: 285-289.

Kimmins, J.P. (Hamish) (1996). The health and integrity of forest ecosystems: Are they threatened by forestry? *Ecosystem Health* 2: 5-18.

Kimmins, H. (1997a) The Peter Pan principle in renewable resource conflicts. Chapter 2 In: *Balancing Act: Environmental Issues in Forestry*, 2nd Edition. UBC Press, Vancouver. pp. 6-13.

Kimmins, J.P. (Hamish) (1997b) Ecology and the ecosystem concept. Chapter 3 In: *Forest Ecology: A Foundation for Sustainable Management,* 2nd Edition. Prentice Hall, Englewood Cliffs, NJ, pp. 23-30.

Kimmins, J.P. (Hamish) (1997c) Biodiversity and its relationship to ecosystem health and integrity. *Forestry Chronicle* 73: 229-232.

Kimmins, H. (1997d) Certification: a market-driven mechanism to promote sustainable forest management. Chapter 18 In: *Balancing Act: Environmental Issues in Forestry,* 2nd Edition. UBC Press, Vancouver, pp. 255-270.

Kimmins, J.P. (Hamish) (1999) Biodiversity, beauty and the "beast": Are beautiful forests sustainable, are sustainable forests beautiful, and is "small" always ecologically desirable? *Forestry Chronicle* 75: 955-958.

Kimmins, J.P. (Hamish) (2000) Respect for nature: an essential foundation for sustainable forest management. In: R.G. D'ion, J. Johnson, and E.A. Ferguson (eds), *Ecosystem Management of Forested Landscapes: Directions and Implementation*. Ecosystem Management of Forested Landscapes Organizing Committee, Vancouver, pp. 3-24.

Kimmins, J.P. (Hamish) and D.M. Duffy (1991) Sustainable forestry in the Fraser River Basin. In: A.H.J. Dorcey (ed) *Perspectives on Sustainable Development in Water Management: Towards Agreement in the Fraser River Basin*. Westwater Research Centre, Faculty of Graduate Studies, University of British Columbia, Vancouver, pp. 217-240.

Kimmins, J.P. (Hamish), D. Mailly and B. Seely (1999a) Modelling forest ecosystem net primary production: the hybrid simulation approach used in FORECAST. *Ecological Modelling* 122: 195-224.

Kimmins, J.P. (Hamish), B. Seely, D. Mailly, K.M. Tsze, K.A. Scoullar, D.W. Andison, and R. Bradley (1999b) FORCEEing and FORECASTing the HORIZON: Hybrid simulation modeling of forest ecosystem sustainability. In: A. Amaro and M. Tome (eds) *Empirical and Process-Based Models for Forest Tree and Stand Growth Simulation*. Edicoes Salamandra, Lisboa, Portugal, pp. 431-441.

Leopold, A. (1953) The land ethic. In: *Round River*. Oxford University Press,

Oxford.

Nassauer, J. (1995) Messy ecosystems, orderly frames. *Landscape Journal* 14: 161-170.

Peterken, G.F. (1996) *Natural Woodland Ecology and Conservation in Northern Temperate Regions*. Cambridge University Press, Cambridge, UK, 522 pp.

Seely, B., J.P. (Hamish) Kimmins, C. Welham and K. Scoullar (1999) Defining stand-level sustainability. Exploring stand-level stewardship. *Journal of Forestry* 97(6): 4-10.

Chapter five:

Why Do You Think that Hillside is Ugly? A Sociological Perspective on Aesthetic Values and Public Attitudes About Forests

David B. Tindall
Department of Forest Resources Management, University of British Columbia
Vancouver, British Columbia

1   Introduction

About five years ago, I attended a public forum on forestry, organized in response to the controversy around Clayoquot Sound. During a number of sessions, members of the audience - who seemed to be the interested, general public - raised issues about clearcutting, many to do with moral and aesthetic values. Another Forestry academic, in addressing the audience, said something to the effect that:

> Forestry is about science. What you are talking about are values. Values are beyond the realm of science. If you don't like the way things are done, go and work to elect a different government when the next election comes around.

There are a great many things that could be said in response to such a statement, but it should suffice to say that the chapters in this volume represent a much broader view about what forestry should be about - that people concerned with forestry planning need to take the public's views into account; and that forestry is not simply *a science* as defined above.

I conduct research on public opinion about forestry issues. This chapter provides commentary on a number of issues relevant to providing a multi-faceted understanding of survey research on aesthetics and sustainability from the perspective of a social scientist. For illustrative purposes, I focus on aesthetic values and their connection to specific attitudes and opinions about forest management. Later I provide a brief empirical example examining the relationship between ratings of aesthetic values and opinions about management of visual characteristics of forests.

2 Clarifying Human Values Associated with Forests

I begin by critically examining the concept of *values*. As a type of short hand, I sometimes refer to an ongoing project of mine as being about *forest values* (see Tindall and Lavallee, 1999). From a social science perspective however, the term *values* is often misused in the context of forestry issues; values are often thought of as physical things in the woods. The next question is usually, how much value does a particular thing have?

As an example, several years ago a speaker at the UBC undergraduate forestry field school raised the question, "What is the value of that wildlife tree?" Those of us present stood around waiting for someone to answer. Eventually he answered his own question by responding that the value of the wildlife tree was $64,000 or some such figure. He then went on to detail the operations costs entailed by leaving and protecting this particular wildlife tree.

While management decisions do have economic costs, this is not generally what social scientists are referring to when they talk about values - at least non-economists. In social science terms, values are cultural ideas about what are desirable goals and what are appropriate standards for judging actions. Put a slightly different way, they are emotionally charged beliefs about what is desirable, right, and appropriate (Rokeach, 1973; 1979; Hagedorn, 1994).

Values are broader in scope than normative beliefs or attitudes - two related concepts. An attitude refers to a general, learned, and relatively enduring tendency on the part of individuals to respond negatively or positively to a given phenomenon. Sociologists tend to view attitudes in relation to social values (*i.e.* as the subjective aspect of values), while psychologists tend to focus on the relationship between attitudes and other aspects of the individual personality. So, in sum, values in this sense are to be located in people's heads, and in the behaviours and artefacts through which these values are manifested. Some human and social values are connected to forest characteristics and forest management, but one cannot simply measure things in the woods to understand human and social values associated with forests.

A good deal of work on *forest values* (as well as on other natural resources and environmental values) is structured around asking respondents "how much are you willing to pay ...?" While this type of research has its uses, what is much more interesting to a social scientist - and ultimately more useful for forest managers and policy makers who wish to understand intergroup conflict over forest management - is to look at the ways in which values underlie positions and concerns about forestry. *Willingness to pay* - a common economic measure - is not really about values, but about valuation in monetary terms. While such analyses have their uses, they do not provide insight into the meaning underlying such preferences, and they are inappropriate for many types of issues. For example, consider the question, "How much are you willing to pay to have women paid the same as men for work of equal value?". While there are some real economic considerations in establishing gender equity, in today's world, the question is beside the point because this is a moral issue. The appropriate response is: just do it!; for further discussion on some of these issues see: Rokeach (1973; 1979), Schwartz (1992), and Foster (1997).

3 Understanding Attitudes/Opinions About Forests and Forest Management

In studying aesthetic and other human/social values using survey methods, as is common in landscape preference studies, one potential pitfall is losing the ability to achieve a deep understanding of people's responses because of the use of standardized formats. Questionnaires and related data collection procedures allow limited opportunities for understanding contingencies. I use the term contingencies in two ways in this chapter. First of all, I use it to refer to the immediate conditions under which an attitude, opinion, or value is elicited. For example, if someone dislikes clearcutting does this mean they have a universal dislike for clearcutting or do they only dislike clearcutting only under certain conditions (clearcutting of coastal rainforests/large clearcuts/clearcutting of old growth)? The second way in which I use the term contingency is to understand how specific attitudes, opinions, and values are linked to other social processes and structures. For example, what relatively general values underlie more specific attitudes and opinions about forestry issues. How are relatively general values related to social structural location (*e.g.* in terms of community, class, and social network positions)?

In discussing these issues, I draw some examples from the Forest Values Project (Tindall and Lavallee, 1999). For this project we conducted 302 face-to-face interviews (averaging about an hour and a half each) with BC residents. Following this, we developed a closed-ended mail-out questionnaire which we then sent back to the initial sample. The questionnaire was 27 pages long, and exhaustive in scope. In the questionnaire, there were a variety of questions on a wide range of values, and on more specific concerns such as visual corridors - I present some of these findings at the end of this chapter.

During the interviews we asked respondents open-ended questions about a wide variety of topics related to the characteristics of forests, and forest management. The goal was to elicit concerns, attitudes, opinions, and underlying values. One of the questions dealt with aspects of the physical environment that people felt were important. In the process of probing for further responses, we often asked people what they thought about *visual corridors*. Responses to this series of probes were very instructive, and in fact led to the development of a series of closed-ended questions on the follow-up questionnaire. However, without having first done these interviews, we would probably have developed some overly simplistic questions about visual corridors that masked underlying concerns.

For the purpose of illustration, it is instructive to recount three interviews that were conducted. In a typical interview with a Ministry of Environment Manager, the informant stated that he understood that some logging had to occur, but that he did not particularly care to stare at clearcuts outside his window, or view them as he drove to work - so he viewed *scenic corridors* in a positive light. Two other respondents viewed scenic corridors negatively, but their underlying concerns diverged considerably. An environmental organization representative I talked to said that he objected to visual corridors because they served to conceal "bad forest practices" and thus were dishonest. By comparison, a logging contractor I interviewed also disliked visual corridors immensely, and he concurred that they were a dishonest practice. His rationale, however, was different. He argued that we should be proud of logging - and stated that he was proud to be a logger. Designing visual corridors was akin to saying we are embarrassed by logging, and we want to hide any evidence of logging. He also said that he truly did not understand why people found old growth forests to be appealing. He much preferred the vitality of a very young forest.

I have provided these examples to illustrate the point that as researchers we must employ research designs that enable us to uncover the meaning underlying survey responses, and that allow for the examination of contingencies. In the Forest Values Project, in the absence of interview responses, we might simply have asked people to rate the desirability of visual corridors; in such an instance we would not have been able to distinguish between the negative responses of the environmental organization representative and the logging contractor.

While I believe that studying the views of the public about general issues such as clearcutting is valid, I nevertheless have some methodological concerns about what it is that people mean when they provide preference ratings to different visual stimuli in research on visual landscapes. In order to articulate these concerns another example from the Forest Values Project (Tindall and Lavallee, 1999) will serve. During interviews with a variety of people - tour guides in particular - respondents would tell me that most tourists and visitors could not tell the difference between old growth and second growth forests. For example, several tour guides in the Queen Charlotte Islands - or *Haida Gwaii* - told stories about taking tourists out on boats, and hearing them express outrage at island hillsides that had obviously been clearcut in recent years. Then as the boats moved along and they passed lush green forested hillsides, they talked about their emotional connections to these vistas and how these old growth forests were clearly spectacular, and distinguishable from a second growth forest. At this point in the interview, the guides would inform me that the tourists were in fact looking at second growth forests but did not realize it! In most cases the guides did not correct their guests for fear of ruining their experience.

So what do we learn from this? The question is, were people in love with the *visual beauty* of the forests they saw and conversely repelled by the *visual ugliness* of the clearcuts, or were they in love with the *idea of old growth forests* and repelled by the *idea of clearcutting*? Research that focuses upon ratings of visual images may not completely address this question. I will return to this issue latter.

4   Studying "The Public"

One methodological issue to consider in survey-based research on forestry issues, is the question, "Who is the public?" While there are a host of issues to consider with regard to this topic, I will briefly address two: the issue of respondent competency, and the definition of community.

One principal of survey research is to ensure that respondents are competent to answer the questions that are asked of them. Conversely, they should be given the opportunity to respond "don't know" or "no opinion" if they truly do not have an informed opinion. One reason for this, is that people will respond to almost anything if given the opportunity. This is one explanation perhaps, why about 13% of the population report that they think Elvis is still alive.

However, this *competency to respond* criterion depends on one's research objective. For example, in a Forestry Canada survey administered to a nation-wide sample, I included a question about clearcutting: whether "it is used too widely", "its level of use is just right", or "it is not used widely enough". I have collected survey data on this question from members of the general public, from members of formal environmental organizations, and also from first year forestry students over three years of classes. The average responses for the first year forestry students generally fall somewhere in between those of the general public, and environmentalists. There are some interesting exceptions - compared to the two other groups, *fewer* forestry students feel that clearcutting is overused as a method. On the other hand, forestry students feel a *larger* proportion of the landbase should be protected as wilderness. In giving guest lectures and public talks, I often present results of these surveys and encourage the audience to discuss them. On occasion, a member of the audience will assert that "the question is biased". The main concern is that I have not asked enough contingency questions. Others comment that the responses to this question are irrelevant, because aside from the forestry students, the samples (meaning the general public and environmentalists) do not know anything about forestry. I presented these survey results, indicating a dissatisfaction with clearcuts, to a fourth year forestry class once, and several disgruntled students said that only the views of the first year forestry students were legitimate (and not those of the general public or environmentalists). However, once findings showing that even the views of the majority of the first year forestry students diverged from their own views were presented, these fourth year students argued that the responses of all three groups should be discounted because they didn't know anything about forestry. Whether or not this claim is true, I counter that these data are still relevant because public perceptions - regardless of the actual state of the forests - are important because they ultimately affect policy-makers and decision-makers.

I turn now to the issue of *who is the public*. Again, to draw from our Forest Values Project (Tindall and Lavallee, 1999), we interviewed most of the District Forest Managers and several of the Regional Forest Managers in the Province of British Columbia. One comment we received repeatedly from these managers was that they knew in intimate detail the concerns, attitudes, and opinions of what they called *the extremes on both sides*: environmentalists, and forest industry advocates. What they did not know they said , is what the *general public* thought about forestry issues. They indicated that this information would be valuable to them.

There has, of course, been a considerable amount of survey research on the views of the general public in British Columbia, some of it good, and some of it not so good. Much of this research has focused, for a variety of reasons, on communities. Often forestry research defines communities by drawing a small circle on a map, and then declaring that whatever is inside of the circle is part of the community. From a social science perspective this is a problem, because it only captures one dimension of the definition of community. Three standard conceptual dimensions of community exist; these include (1) shared geography; (2) collective identity; and (3) structural integration (Wellman, 1979). While these three dimensions of community are interrelated, they can be analytically separated and treated as semi-autonomous indicators of *community*.

Shared residential location is probably the most frequently used *geographical* definition of community. This definition is used by geographers, sociologists, and planners alike.

Collective identity is a cognitive aspect of community that refers to the way in which people identify themselves with some larger social group. Identification can be made both about who belongs to the group, and who are outsiders. A related cognitive aspect of community is "*sentiments of solidarity*" (Wellman, 1979). While identification refers to the cognitive content of group affiliation, solidarity refer to its emotional intensity. In other words, sentiments of solidarity refer to the existence of strong emotional ties to the group. Although, analytically part of the cognitive domain, this concept is sometimes conceptualized as part of structural integration.

Structural integration, the final concept to be considered here, refers to patterned interactions amongst *community members*. These patterned interactions can take a number of forms. Many social scientists utilize social networks to depict the structural dimension of communities (Tindall *et al.*, 1999). Analysts who focus on community as shared geography or neighbourhood frequently overlook, or assume the *identity* and *social structure* parts of this definition - and hence, neglect to obtain empirical data on these aspects of community.

Substantively, the use of geographically-based definitions of community is problematic in a province such as British Columbia for two reasons. The most obvious is that most forested land in BC is crown land and therefore owned by the general public of the province, not just by local communities. Secondly, while there are local communities in geographic terms, there are also spatially dispersed communities of forest users. Thus, capturing these different communities provides a methodological challenge. In some cases collecting data on local, geographic communities is appropriate. In many other cases it is not. Methodological sophistication in terms of sampling is required.

In considering the design and interpretation of research to illuminate people's landscape perception, there is the need to capture intergroup (and intra-group) differences. Again, in forestry, there is a tendency to talk about the general public as if there exists a single homogeneous entity. More emphasis needs to be given in research exploring aesthetics and sustainability on intergroup differences as indicated by a number of striking patterns of inter-group differences in values, attitudes, and opinions obtained by the Forest Values Study (Tindall and Lavallee, 1999).

There are some obvious types of inter-group comparisons to be made: for example, differences between different economic sectors, between urban and rural residents, and between those with varying degrees of formal education. There are also some other intergroup differences that may not seem so obvious, or that present special challenges. One example is gender. Survey research has revealed persistent attitudinal and opinion differences on a variety of environmental issues between men and women. While there are some issues where gender differences do not seem to occur, where there are differences, women are usually more concerned about environmental problems than are men (Brody, 1984; Hamilton, 1985; Blocker and Eckberg, 1989). This finding is particularly relevant in the present context given the fact that forestry is a male-dominated occupation.

An example of a set of potential intergroup differences that present a special methodological challenge, is that of examining differences between Aboriginals and non-Aboriginals. In particular, some of the methods that work well with the general non-Aboriginal population (such as self-administered questionnaires or telephone interviews) do not work very well with Aboriginals. One barrier is simply culture, and a lack of familiarity with survey techniques that leads to high rates of non-response. A second issue is the history of cultural appropriation (in particular, by university-based researchers) that makes some Aboriginals reticent to share their knowledge. A third issue in BC is the current state of the Indian Land Question. The fact that most aboriginal groups are involved in the treaty process or land claims process in some way, means that many individuals are reluctant to provide information to survey researchers for fear that the information will be used against their community - by the government, by forest companies, or other parties (Tindall and Lavallee, 1999).

5   Science, Values and Framing Issues: Why the Focus on Clearcutting?

In reference to survey research on public perceptions of forest management, are we getting it right? In trying to understand values, attitudes, and opinions about forestry, are we focusing on the right things? A central concern of scientific measurement is validity; that is, are you measuring what it is you think you are measuring? Can we take what people tell us at face value? In exploring these questions, this section will consider the campaign of the environmental movement against clearcut logging in British Columbia.

In BC there has been a critical pre-occupation with clearcut logging. Some responses to criticisms about clearcut logging have focused on scientific issues. In my opinion, this is probably barking up the wrong tree. There are several reasons for this view: (1) one, because in the words of Stephen Yearly (1992a; 1992b), environmentalists have an ambivalent relationship to science; and (2) because the focus on clearcutting is not based primarily on scientific concerns, it is a communication strategy of the environmental movement.

In discussing the ambivalent relationship that environmentalists have to science, Yearly (1992a; 1992b) points out that science and scientists have uncovered and publicized many of the key environmental problems that concern environmentalists. For example, without modern science, no-one would know about the ozone layer, much less about the well-publicized holes in it. Applied science transported humans into outer space, where we developed a new image of the planet - a blue ball in space, an interconnected biosphere - and this new image helped to foster *Gaia Hypothesis*. On the other hand, applied science has brought us many disasters and potential disasters - from DDT to nuclear bombs.

From the perspective of environmentalists, therefore, science and scientists are unreliable allies. In environmental controversies, one will find scientists for hire working on both sides of an issue. While in some ways radical, in many ways science is inherently conservative. Empirical evidence must be obtained and analysed - often a painfully slow and uncertain process. Also, while many environmentalists are scientists, and the concerns of many environmentalists are rooted in science, the concerns of environmentalists (and others for that matter) tend to be multi-faceted - and not limited to scientific issues.

One of the reasons why environmental issues are often framed in scientific terms is because science has such high esteem in Western culture. It has strong currency. However, many environmentalists are also motivated by their adherence to other values, such as those related to ethics, spirituality, and aesthetics. These types of values do not have as much currency in popular culture, and therefore environmental issues tend to be framed primarily in terms of science. This is why countering criticisms about clearcutting in purely scientific terms - as some of my colleagues have attempted - in my opinion, is bound to be unsuccessful; to a certain extent the debate is really about something else.

The phenomenon referred to in the social movements literature as *framing* (Snow *et al.*, 1986; Snow and Benford, 1988) is worthy of further consideration. The analysis of *framing* or *frame alignment processes*, involves examining the linkages between the perceptions, values and interests of individuals and the recruitment strategies employed by social movement organizations. As described by Snow *et al.* (1986), a key argument is that "frame alignment" is a necessary prerequisite to movement participation:

> By frame alignment, we refer to the linkage of individual and SMO (social movement organization) interpretations, such that some set of individual interests, values and beliefs and SMO activities, goals and ideologies are congruent and complementary. (Snow *et al.*, 1986, p. 464)

The study of framing examines the form that arguments take, and the functions of various types of communication strategies in putting across the argument. The concept of *frame amplification* in social science refers to the process of making salient the key values and beliefs important to a social movement's goals and activities. Proponents of a framing approach argue that while these key values and beliefs may be held by individuals, they exist alongside a whole host of other values and beliefs and thus their importance is relatively ambiguous to potential movement adherents. Support for and participation in movement activities is often dependent on the clarification and reinvigoration of an interpretive frame. An example of frame amplification drawn from a study of the BC environmental movement (Tindall and Begoray, 1993) can be seen in the manner in which a conflict over the Carmanah Valley in British Columbia was handled. Although environmentalists wanted to protect the biodiversity of the ecosystem, this is a difficult concept to communicate to the public. However, as one activist stated, it was easier for the public and the press to grasp that the Carmanah has:

> Canada's largest tallest trees - that's a sexy thing. The press - the CBC loves it, cause they can just go in, take a picture of the tallest trees, and walk out, and say they're going to be cut. That's why Carmanah was saved - at least fifty percent of it ... that's why Carmanah worked. (environmental activist)

Another concept that is relevant here is what is called "*frame resonance*". The basic idea is that arguments that are framed to resonate, or strike a responsive chord with aspects of the dominant culture and experience of individuals, are more likely to be successful than those that do not. The instance of Carmanah reflects the fact that *big* things are prized in North American society, as in the expression "bigger is better".

The mode of communication is also important. As psychologists and others have noted, humans are very much visual beings. Environmentalists have packaged their anti-clearcut campaign for visual media very effectively. Focusing on visual images of clearcuts is another example of frame amplification. While these arguments are well-suited for brief news-stories on television, postcards and posters have been utilized to show dramatic images of clearcuts, and these images find their way into the newsprint media. In a content analysis of print media coverage of the conflict over forestry and conservation in British Columbia between 1988 and 1992 in British Columbia, "Clearcut" was the second most frequently appearing keyword, appearing in 25% of all stories (Tindall and Doyle, 1997). The forest industry also uses various framing techniques; particularly successful are those that focus on job loss.

Some things are easier to frame than others. Clearcutting is easier to frame than biodiversity. Seal pups and bear cubs have more *cultural resonance* as being cute and cuddly than codfish do. I have never seen a stuffed codfish in Toys R Us. While there is much more that could be said about the focus on clearcutting as a communication strategy, my point is, in doing research on the aesthetic characteristics of landscapes, we need to remember that concerns about clearcutting are not based solely on direct, unmediated, perception of the landscape.

6   The Link Between Values and Attitudes/Opinions: Survey Results on Visual Issues

The final section of this chapter briefly describes some survey results from the Forest Values Project (Tindall and Lavallee, 1999), examining the relationship between ratings of the importance of aesthetic values and opinions about management of visual forest characteristics. The data described were obtained from a quota sample of British Columbia from all six forest regions of the province. The sampling strategy was developed to ensure variation in terms of people's relationship to the forest. Sampling criteria included: forest region, community, gender, sector of employment, and occupation. More details about the methodology are provided in Tindall and Lavallee (1999).

Table 5.1 is a cross-classification table describing the relationship between respondent's scores on an aesthetic values index (see below), and their opinion about a variety of visual forest management issues. The purpose of this analysis is to illustrate how general values underlie preferences about more specific visual forest management preferences.

The forest values questionnaire (Tindall and Lavallee, 1999) contained 10 different sections. The first section of the questionnaire required respondents to rate a series of value indicators associated with 12 value categories, one of which was aesthetics. For example, one indicator for aesthetic values was: "Please indicate the importance of ..."; "The beauty of natural areas surrounding your community. To me this value is ... (1) Not Important, (2) Somewhat Important, (3) Very Important, (4) Extremely Important". The indicators associated with *aesthetic values* in the survey results for the total sample combined to produce an alpha score = 0.84, indicating strong inter-item reliability. An index was created from these indicators. To interpret the index score: the higher the index score, the more important aesthetic values were to the respondent. For the purpose of providing an easily interpretable example, I have transformed the continuous *aesthetic values index* into a dichotomous variable. The categories are *Low* (all values at or below the median) and *High* (all values above the median). The opinion items were developed directly from interview data, and reflect sentiments articulated by respondents.

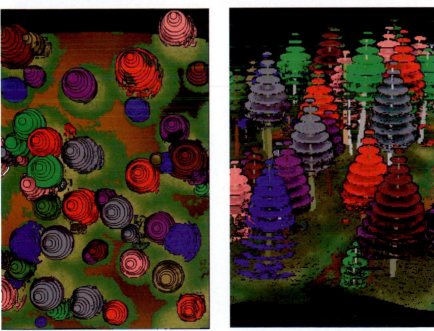

Plate 1. Examples of the FORCEE stand-level ecosystem model interface. (Kimmins, UBC.)

Plate 2. Sitka spruce, Lyell Island, Queen Charlottes. Beautiful Sitka spruce that requires periodic stand-replacing disturbance. The beautiful mossy glade condition is unnatural and an artefact of introduction of deer by European missionaries in the late 1800s without their natural predators. Beautiful, but unnatural and unsustainable.

Plate 3. Seeing is not believing. Ugly clearcut on west coast of Vancouver Island. Old growth values destroyed, but ecosystem remains productive and vibrant. Rapidly growing young climax rainforest.

Plate 4. Van Gogh, *The Harvest.*

Plate 5. Church, *Morning in the Tropics.*

Plate 6. The farmhouse as a visible symbol of longstanding stewardship. (D. Cavens, Landscape Architecture, UBC.)

Plate 7. Heavy concentrations of downed wood usually attract adverse aesthetic ratings in perception experiments. (BCMoF.)

Plate 8. Geometric clearcuts seen in middleground symbolize insensitivity to the local landscape, but demonstrate a high degree of coherence and orderliness. (Sheppard.)

Plate 9. Visualization of proposed forest harvest patterns in northern Ontario, Canada. (Orland, IMLAB.)

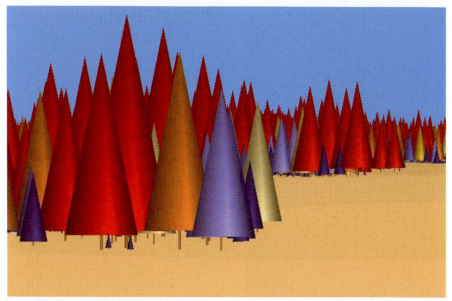

Plate 10. Tree canopies coloured by trunk diameter (diameter at breast height).

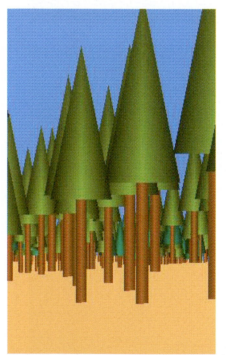

Plate 11. SmartForest views. Left: iconographic (management mode). Right: realistically textured (landscape mode).

Plate 12. Synthetic image of unbuilt resort proposal by Frank Lloyd Wright for Emerald Bay, Lake Tahoe. (CLR, U of T, Canadian Centre for Architecture. Synthesized using CLR's *PolyTRIM* software.)

Plate 13. Illustration of overall air pollution in the Rhine Valley in the Grisons, Switzerland. (ORL Institute, ETHZ, CLR, U of T. Synthesized using CLR's *PolyTRIM* software.)

Plate 14. Image processing photo simulation of proposed strip-harvesting at Gavin Lake in British Columbia. (John Lewis and Stephen Sheppard, Centre for Advanced Landscape Planning, UBC, Vancouver.)

Plate 15. Synthetic computed simulation of proposed strip-harvesting at Gavin Lake in British Columbia using World Construction Set. (John Lewis and Stephen Sheppard, Centre for Advanced Landscape Planning, UBC, Vancouver.)

Scanned print film photograph.

Simulation using Vistapro with overlaid ground colours.

Simulation using ERDAS Imagine.

Simulation using Vistapro with automated terrain coloration and manual object placement.

Plate 16. Landscape visualizations, Linn of Corriemulzie.

Plate 17. Giant Sitka spruce, Windy Bay, Lyell Island, Queen Charlottes. Unknown age. Unsustainable without periodic disturbance by wind.

Plate 18. West Coast old growth. Carmanah Valley. Climax northern temperate rainforest. Beautiful, natural and sustainable.

Plate 19. Seeing is believing. Excessive levels of soil disturbance and compaction caused by skidder logging on steep slopes in southeastern British Columbia.

Plate 20. An automatic snow generation algorithm can be used in scenes where the depth of the snow is visually important. In order to produce a 3D model of the snow cover, the computer computes the location of the fallen snow and detects and resolves any resulting avalanches. (P. Fearing, Computer Science, UBC.)

Plate 21. An example of multi-scale snow generation. The haystack consists of thousands of tiny components, which must be resolved simultaneously with the much larger surrounding field. (P. Fearing, Computer Science, UBC.)

Table 5.1. Cross-classification of aesthetic values index by ratings of visual forest management opinion items (in percentage).

Opinion items	Respondent's Score on Aesthetic Values Index (High or Low)	Ratings in Percentage								χ^2 Value and Significance
		Other	Strongly Agree	Mostly Agree	Partly Agree/ Disagree	Mostly Disagree	Strongly Disagree	Don't Know/ No Opinion	N	
1. Clearcutting should not be stopped just because people think it is ugly.	High	2.4	15.5	28.6	16.7	31.1	22.6	1.2	84	χ^2 = 22.20, p ≤ 0.001
	Low	0	37.6	34.1	16.5	2.4	9.4	0	85	
2. Visual corridors should not be created just because people think logging is ugly.	High	2.4	10.6	20.0	20.0	28.2	15.3	3.5	85	χ^2 = 18.14, p ≤ 0.01
	Low	0	23.5	29.6	25.9	16.0	4.9	0	81	
3. The creation of visual corridors should be discouraged - because they give a false impression of forest management by covering up the bad forest practices of forest companies.	High	1.2	8.3	14.3	31.0	31.0	11.9	2.4	84	Non-significant
	Low	0	8.6	23.5	30.9	25.9	11.1	0	81	
4. Visual corridors are unnecessary - we should take pride in logging instead of trying to hide it.	High	2.4	7.1	5.9	16.5	32.9	29.4	5.9	85	χ^2 = 27.57, p ≤ 0.001
	Low	0	13.6	17.3	30.9	30.9	7.4	0	81	
5. Visual corridors are good for tourism.	High	0	20.0	50.6	12.9	3.5	4.7	8.2	85	χ^2 = 21.66, p ≤ 0.001
	Low	1.3	15.2	41.8	32.9	8.9	0	0	79	
6. More attention should be devoted to minimizing the visual effects of logging on the landscape.	High	0	28.2	32.9	24.7	9.4	1.2	3.5	85	χ^2 = 28.60, p ≤ 0.001
	Low	0	4.9	25.9	29.6	32.1	6.2	1.2	81	

In panels 1 to 6 of Table 5.1, I provide a series of cross tabulations that examine the bivariate relationship between the aesthetic values index and a series of opinion items (*panel* refers to the rows corresponding to the opinion item on the left. The term panel is used as there are two rows corresponding to each of the opinion items). For example panel 1 relates to the opinion item, "Clearcutting should not be stopped just because people think it is ugly." Of people who highly rated aesthetic values, 15.5% strongly agreed with the statement. By contrast, of respondents with a low rating for aesthetic values, 37.6% strongly agreed with the statement.

In all instances but one (Item 3), there was a statistically significant relationship in the direction we would expect based on knowledge of respondents' ratings of aesthetic values. By the expected direction we mean, people who have a greater preference for aesthetic values are more likely to favour ameliorating the visual effects of clearcutting, to favour the creation of visual corridors, etc. (In the exception, the relationship was not statistically significant but still in the same direction.)

In panel 1 of Table 5.1 we observe that respondents who rated the importance of aesthetic values low, were more likely to *strongly agree* or *mostly agree* with the statement "clearcutting should not be stopped just because we think it is ugly" ($\chi^2 = 22.20$, p $\leq$ 0.001). In panel 2 of Table 5.1 we observe that respondents who rated the importance of aesthetic values low, were more likely to *strongly agree* or *mostly agree* with the statement "visual corridors should not be created just because people think logging is ugly" ($\chi^2 = 18.14$, p $\leq$ 0.01).

Panel 3 examines the relationship between the aesthetics values index and opinions regarding the statement "the creation of visual corridors should be discouraged because they give a false impression of forest management by covering up the bad forest practices of forest companies". For this panel, there is no statistically significant association between the two variables, though the relationship is still in the same direction (*i.e.* those who rated aesthetic values low were more likely to *mostly agree* with the statement, while those who rated aesthetic values high were more likely to *mostly disagree* with the statement). There are several possible reasons for why this item was not statistically significant - one being the *double-barrelled* nature of the opinion item. Different sub-groups of people may be agreeing or disagreeing with different parts of the statement - and thus cancelling each other out.

Panel 4 of Table 5.1 demonstrates that respondents who rated the importance of aesthetic values high, were more likely to *strongly disagree* or *mostly disagree* with the statement "visual corridors are unnecessary - we should take pride in logging instead of trying to hide it" ($\chi^2 = 27.57$, p $\leq$ 0.001).

Panel 5 of Table 5.1 shows that respondents who rated the importance of visual/aesthetic values high, were more likely to *strongly agree* or *mostly agree* with the statement "visual corridors are good for tourism" ($\chi^2 = 21.66$, p $\leq$ 0.001).

Results in panel 6 of Table 5.1 reveal that respondents who rated the importance of aesthetic values high, were more likely to *strongly agree* or *mostly agree* with the statement "more attention should be devoted to minimizing the visual effects of logging on the landscape" ($\chi^2 = 28.60$, p $\leq$ 0.001).

As noted above, this pattern of findings is consistent with expectations, and highlights the importance of examining the values that underlie opinions and attitudes about forest practices, and forest land use. This particular example provides a relatively simple illustration of a theme touched upon earlier. A more thorough analysis of these items would involve multivariate analysis techniques (such as multiple regression) and include multiple values (*e.g.* perhaps ecological and economic value indexes in addition to the aesthetic values index) to predict opinion/attitude items.

7 Conclusion

The title of the workshop for which the articles in this volume were originally presented was, *Linking Sustainability to Aesthetics: Do People Prefer Sustainable Landscapes?* The answer to that question is, r = 0.53, p < 0.001. This is the zero order correlation between the ecological values index and the aesthetic values index. Thus, at least in a social psychological sense, the answer to the workshop question is "yes". Of course much more can be said about this finding - but I will leave it for another article.

In this chapter, I have explored the issue of understanding preferences for and the linkages between sustainability and aesthetics from a social science perspective. In so doing, I have questioned the validity of studies that simply ask respondents to rate visual images. I have also attempted to highlight a variety of methodological obstacles that survey researchers and interpreters need to be aware of. Understanding the nature of human values associated with sustainability and with aesthetic preferences is a complex undertaking, but an improved understanding, will lead to better forest management decisions and policies.

Acknowledgements

I would like to acknowledge Loraine Lavallee's contribution as a collaborator on the Forest Values Project.

References

Blocker, T.J. and D.L. Eckberg (1989) Environmental issues as women's issues: General concerns and local hazards. *Social Science Quarterly* 70(3): 586-593.

Brody, C. (1984) Differences by sex in support for nuclear power. *Social Forces* 63(1): 209-228.

Foster, J. (ed) (1997) *Valuing Nature? Economics, Ethics and Environment*. Routledge, New York.

Hagedorn, R. (ed) (1994) *Sociology*. 5th edition. Harcourt Brace and Company, Toronto.

Hamilton, L. (1985) Concern about toxic wastes: Three demographic predictors. *Sociological Perspectives* 28(4): 463-486.

Rokeach, M. (1973) *The Nature of Human Values*. Free Press, New York.

Rokeach, M. (ed) (1979) *Understanding Human Values: Individual and Societal*. The Free Press, New York.

Schwartz, S.H. (1992) Universals in the content and structure of values: Theoretical advances and empirical tests in 20 countries. In M. Zanna (ed), *Advances in Experimental Social Psychology* (Vol. 25, pp. 1965). Academic, Orlando, pp. 1-65.

Snow, D.A. and R.D. Benford (1988) Ideology, frame resonance, and participant mobilization. In: B. Klandermans, H. Kriesi, and S. Tarrow (eds), *From Structure to Action*. JAI Press, Greenwich, pp. 197-218.

Snow, D.A., E. Burke, S.W. Rochford and R.D. Benford (1986) Frame alignment processes, micromobilization, and movement participation. *American Sociological Review* 51: 464-481.

Tindall, D.B. and N. Begoray (1993) Old growth defenders: The battle for the Carmanah Valley. In: S. Lerner (ed), *Environmental Stewardship: Studies in Active Earthkeeping*. Waterloo, Ontario: University of Waterloo Geography Series, pp. 269-322.

Tindall, D.B. and A. Doyle (1997) *Wood frames: An examination of environmental movement framing of forestry and conservation in British Columbia*. Paper presented at the Annual Meetings of the American Sociological Association. Toronto, Ontario, August, 1997.

Tindall, D.B. and L. Lavallee (1999) *A Report on the Forest Values Questionnaire: Development, Administration, and Evaluation*. Department of Forest Resources Management, University of British Columbia.

Tindall, D.B., F.M. Kay, and K.L. Bates (1999). Urban and community studies. In: L.R. Kurtz (ed), *Encyclopedia of Violence, Peace, and Conflict, Volume 3*. Academic Press, San Diego, pp. 603-624.

Wellman, B. (1979) The community question. *American Journal of Sociology* 84: 1201-1231.

Yearly, S. (1992a) Green ambivalence about science: Legal-rational authority and the scientific legitimation of a social movement. *British Journal of Sociology* 43(4): 511-532.

Yearly, S. (1992b) *The Green Case: A Sociology of Environmental Issues, Arguments and Politics*. Routledge, London.

PART III

Perspectives on
Forest Sustainability

Chapter six:

Criteria and Indicators of Sustainable Forestry: A Systems Approach

Chadwick D. Oliver
College of Forest Resources, University of Washington
Seattle, Washington

J.P. (Hamish) Kimmins, Howard W. Harshaw,
and Stephen R.J. Sheppard
Faculty of Forestry, University of British Columbia
Vancouver, British Columbia

1   Introduction

One of the reasons why European countries prospered compared to their Middle Eastern and Asian counterparts during the early Middle Ages was because the Europeans had a relatively orderly process for the transfer of power upon the death of a monarch. With no orderly process, the death of a leader in the other countries resulted in civil wars (Maalouf, 1984). The orderly management of the transfer of power has by now been instituted in many of the world's political systems, so that change in leaders occurs with little chaos - and the values that they espoused are maintained. Recently, the world's economic system has become relatively stable, though still dynamic, because an orderly process of adjusting to financial changes has been institutionalized. In contrast, over the past two decades the processes for changing the values to be provided by forests have proven to be relatively ineffective at providing an orderly transition, as evidenced by the extreme rancour generated over forestry issues. More turmoil has resulted in the United States from the change in forest values than with the changing of Presidents. The result has been a reduction of many traditional values provided by forests (*e.g.* timber and employment) with very little evidence of actual enhancement of the desired new values (*e.g.* biodiversity and protection of endangered species). The irony is that most people in the United States probably agree on the values to be provided by forests; the disagreement is less about *what values to provide* than about *how to provide them*.

Before any values can be provided consistently from forests, an orderly process is needed determine how to provide them - whatever they may be. There is a general consensus that the world is or can be wealthy enough to be concerned with future generations; this consensus is expressed as *sustainable development* (Brundtland, 1987):

> development that meets the needs of the present without compromising the ability of future generations to meet their own needs.

In an application of this principle to forested landscapes, the Canadian Council of Forest Ministers (1998) defined sustainable forest management as:

> ecologically sound practices that maintain the forest ecosystems' integrity, productivity, resilience and biodiversity. That involves sustaining a wide range of ecological processes through which plants, animals, micro-organisms, soil, water and air interact.

As will be discussed in this chapter, the broad principle of sustainable development is being sub-divided into more specific objectives or *criteria* for sustainable forestry that are informed and measured by specific *indicators* in order to provide unambiguous targets for management (see Table 6.1). A process that allows these objectives and indicators to be achieved (and changed when necessary) without rancour can be instituted by regarding the forest as a system - an *ecosystem* - and through the application of the principles of *systems management.*

Table 6.1. Criteria for the Conservation and Sustainable Management of Temperate and Boreal Forests developed through the Montreal Process. These criteria are being translated to more specific indicators.

The Montreal Process - Criteria for Sustainable Forestry
1. Biological diversity.
2. Productive capacity.
3. Forest health and vitality.
4. Soil and water conservation.
5. Contribute to global carbon sequestration.
6. Socio-economic benefits.
7. Legal, institutional, and economic frameworks.

This chapter first describes a systems approach to the hierarchical organization of ecosystem management processes (Section 2); it then examines how such an approach might work in organizing criteria for evaluating sustainability across geographic scales and as an aid to decision-making in ecosystem management (Section 3); lastly, some of the particular criteria and indicators of sustainable forest management which would be suitable for inclusion in a hierarchical system are described (Section 4).

2   A Systems Approach to Forest Management

In the past, forests had been thought of as steady-state systems. Barring *unnatural* human disturbances, each forest stand was assumed to contain a permanent group of species which maintained themselves in an equilibrium of birth (or regeneration) and death. Forest products and other values could be provided by selective harvesting, which allegedly mimicked the natural process. Management according to this perspective required little co-ordination over time or among stands, since each stand was a relatively *closed* system. There is now recognition that stands change over time, that different species depend on different conditions provided by the changes (Botkin and Sobel, 1975; Sprugel, 1991), and that products and other values are best provided by managing in concert with these changes. To provide all species and other human values, the changing stand conditions need to be co-ordinated among stands, at different spatial scales, and through time (Oliver, 1992; Oliver and Larson, 1996). Clearly, this is a complex undertaking.

In the search for ways to avoid being overwhelmed by this complexity, a *systems* approach has emerged in the natural and management sciences as perhaps the best method of analysis and management of the many complex interactions that occur in forests (Oliver and Twery, in press). This approach groups the interactions of *ecosystem components* (*i.e.* activities, functions, or processes) into *modules* (*i.e.* groups or subsystems) and concentrates on the relations among groups. The groups are related through *outputs* from one group flowing as *inputs* to another group. These inputs and outputs (*i.e.* events, structures, patterns, measurable criteria, objectives, or indicators) are also grouped or generalized to avoid complexity. A key to a successful system is to identify the fewest and simplest input/output measures which, as output, would characterise the important interactions occurring in a module and which, as input, would characterize the input needs of another module (or modules).

Systems can be grouped, studied, and managed at many levels - from subatomic systems to the universe as a system. Forest systems, or *ecosystems*, are often grouped into a hierarchy of spatial scales from individual trees to stands, to landscapes, to subregions and regions, to broad forest types, to all forests within a country, or to all of the world's forests (Oliver *et al.,* in press). One of the principal challenges of establishing sustainable forestry lies in the problem that many criteria and indicators are conceived at national and international scales, but the evaluation of good stewardship and sustainable forestry must be made at the stand, local landscape, and regional levels. The hierarchical systems approach suggested here provides a possible mechanism for linking indicators of stewardship at the stand level to indicators of the state of millions of hectares, or of whole countries.

Over one hundred years of management science have provided relatively effective procedures by which to make good decisions. These procedures can be divided into *decision analysis*, *decision-making*, and *implementation*. Decision analysis involves several steps:

- scoping of existing conditions and management issues;
- identifying management objectives or criteria (and developing indicators as measures of whether an objective is being achieved);
- developing alternative means of achieving the objectives; and
- determining the consequences of each alternative for each objective.

Decision-making involves selecting the alternative that provides the desired, or most acceptable, mixture of outcomes and tradeoffs. Decision implementation involves carrying out the alternative choice and documenting the consequences of that management decision through monitoring and feedback. When managing forests or other hierarchical systems that integrate biological and management processes at different spatial scales, the decision procedures constitute inputs/ outputs that flow among the modules at different levels (Figure 6.1). Each level processes the input information and provides output in a suitable form to appropriate modules at the next higher or lower level. The flow of information among hierarchical levels is determined by the general behaviour of systems with sequential processes. Management often entails overcoming *bottlenecks* - modules where processing is delayed for various reasons. A common cause of these bottlenecks is the processing of too much information. To avoid this, input information to and from any module needs to be communicated in a few, simple measures.

3   Organization of Criteria and Indicators of Sustainability

The potential benefits of a systems approach to sustainable forest management is that it offers an orderly and explicit method of integrating factors across the range of geographic scales, from the international to the local, and over time. How might a hierarchical systems approach be applied to the complex task of evaluating the sustainability of forest management or deciding between alternative forest management strategies? This section examines first the application of such a system to the comparative evaluation of the sustainability of forest management across countries and regions, using a hypothetical range of sustainability criteria and indicators. At a more local forest level, the systems approach is then related to the problem of decision-making within forest management strategies that seek to maintain a range of values over time as stand conditions change.

3.1 Comparisons of Forest Sustainability Across Geographic Scales

An example of how *measurable criteria* may be developed for a management objective (or sustainability criterion) at different levels of a national or global organizational hierarchy is described below. It should be remembered that this example is hypothetical, and bears no resemblance to a specific country. Also, the specific indicators suggested here are used primarily as illustrations for discussion

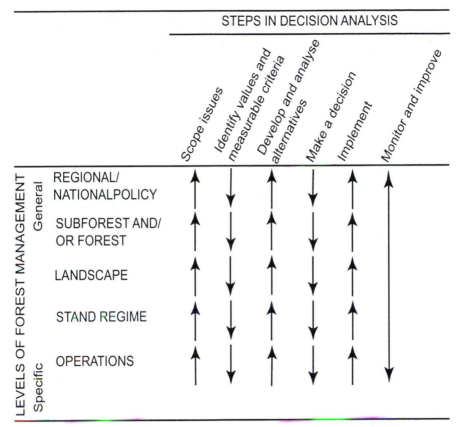

Fig. 6.1. In a hierarchical system such as forest management, there is a sequence of activities in the decision process that occur at different levels. The decision-making activities at each hierarchical level is stimulated by the input of information from above or below, depending on the activity. The direction of flow of information for each activity is shown by arrows. Information is amalgamated upward, and expanded downward to avoid bottlenecks caused by any level processing too much information (from Oliver *et al.*, in press).

purposes, and would need to be examined further in some depth before being applied in earnest. The example will begin at the *international* hierarchical level and be expanded to more site-specific levels. Input of information would in fact occur in the opposite direction, beginning at the more site-specific levels. The example assumes that criteria (drawn from the Montreal Process, as described below) are to be compared among countries (see Table 6.2).

In this application, it is necessary to compare relatively simple numbers to determine how well the management of forests in different countries is addressing the many measures of biological diversity at various spatial scales, their

Table 6.2. Example of how robust, measurable criteria of current conditions of forests would be viewed by policy makers at the national and international policy level relative to the various criteria and indicators. Numbers and countries are hypothetical and values are normalized to a maximum of *10*.

	Country A	Country B	Country C	Country X
Criterion 1: Conservation of Biological Diversity				
Indicator 1.1: Ecosystem Diversity				
a) Relative area by forest type	8	8	4	---
b) Forest type area by structure	9	4	6	---
c) Forest type area protected	3	7	7	---
d)				
Indicator 1.2: Species Diversity				
a) Number of species	3	8	7	---
b) Viability status of species	7	3	6	---
Indicator 1.3: Genetic Diversity				
a)	---	---	---	---
Criterion 2: Maintenance of Productive Capacity				

sustained productivity, and other measures of (or values provided by) sustainable forestry. A score of *10* would indicate that this value is being provided (or the sustainability objective is being met) to the maximum needed extent. In examining the hypothetical numbers in Table 6.2, note that each country is providing some values quite well (high scores), and others quite poorly (low scores). The challenge is to develop policies that will improve the scores.

The indicators of biological diversity such as those illustrated in Table 6.2, should result from processing of more complex inputs from lower hierarchical levels (Figure 6.2). Each number is normalized, with the maximum possible value of biodiversity assigned the score of *10*. For example, to obtain measures of *Criterion 1.1a*, "Relative area by type", each nation's forests are divided into ecological types (such as the United States' ecological types, as used in this example), and each forest type is assessed as to whether or not it contains enough area to sustain biological diversity (Table 6.3). Information about each forest type is provided by a lower module in the hierarchy that concentrates on each forest type. For the purposes of this illustration, a forest type contains enough area if it meets two tests:

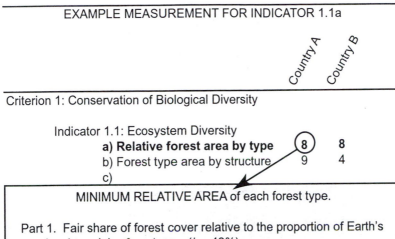

Fig. 6.2. Example of type of information used to develop values for *Indicator 1.1a, Country A* (Table 6.2).

- It must still contain its *fair share* of its original forest cover. The world's forests have been reduced to 43% of their former area during the past 4,000 years (FAO, 1993). Therefore, applying this global average locally, a *fair share* would be 43% of its original forest area. The information on present and former area is collected and passed *upward* from lower hierarchical levels.
- It must contain a *minimum viable area*. It is assumed for the purposes of this illustration that a forest type needs to cover 100,000 hectares to be considered minimally viable (although in fact this may need to be modified for situations outside North America).

In Table 6.3, the score of 8 for *Country A's Criterion 1.1a* indicator (Relative area by type), means that 16 of the 20 forest types meet both these two tests - and that four types do not meet at least one of the tests for various reasons.

As another example, to obtain measures of *Criterion 1.1b*, "Forest type area by structure", each forest type is assessed as to whether or not it contains a minimum amount of each structure needed by native species (Figure 6.3 and Table 6.4). Different forests have different structure needs. However, it is assumed in this example that the maximum needed is five structures to sustain native species (Oliver and Larson, 1996; see Figure 6.3) and that the minimum threshold required for each structure is judged to be 10% of the type's total area. The score of *9* for

Table 6.3. Calculation of relative forest area by type (*Indicator 1.1a, Country A -* see Figure 6.2).

	PART 1 Fair share of forest cover relative to proportion of Earth's previous/remaining forest area.			PART 2 Minimum viable area of forest type (in thousands of ha).			Summary
	Previous/ Remaining	Target Minimum [1]	Acceptable?	Forest Area	Target Minimum [2]	Acceptable?	Accept Both?
EASTERN REGION							
White-Red-Jack Pine	36%	43%	Y	271	100	Y	Y
Spruce-Fir	44%	43%	Y	1 483	100	Y	Y
Longleaf-Slash Pine	45%	43%	Y	124	100	Y	Y
Loblolly-Shortleaf Pine	43%	43%	Y	142	100	Y	Y
Oak-Pine	52%	43%	Y	180	100	Y	Y
Oak-Hickory	56%	43%	Y	1 970	100	Y	Y
Oak-Gum-Cypress	50%	43%	Y	1 012	100	Y	Y
Elm-Ash-Cotton Wood	29%	43%	N	404	100	Y	N
Maple-Beech-Birch	56%	43%	Y	1 014	100	Y	Y
Aspen-Birch	38%	43%	N	481	100	Y	N
WESTERN REGION							
Douglas-Fir	58%	43%	Y	3 080	100	Y	Y
Ponderosa Pine	62%	43%	Y	2 774	100	Y	Y
Western White Pine	24%	43%	N	49	100	N	N
Fir-Spruce	67%	43%	Y	4 720	100	Y	Y
Hemlock-Sitka Spruce	61%	43%	Y	423	100	Y	Y
Larch	58%	43%	Y	155	100	Y	Y
Lodgepole Pine	58%	43%	Y	3 321	100	Y	Y
Redwood	49%	43%	Y	50	100	N	N
Other Western Softwoods	81%	43%	Y	1 595	100	Y	Y
Western Hardwoods	53%	43%	Y	1 945	100	Y	Y
Total:							**16/20**
Normalized Value (0-10)							**8**

[1] $\dfrac{\text{World Forest Area in 1995 AD}}{\text{World Forest Area in 2000 BC}} = \dfrac{3.45 \text{ billion ha}}{8 \text{ billion ha}} = 43\%$

[2] Minimum absolute viable area assumed to be of 100,000 hectares.

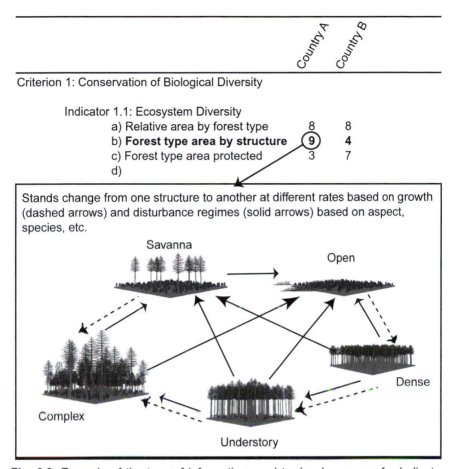

	Country A	Country B
Criterion 1: Conservation of Biological Diversity		
Indicator 1.1: Ecosystem Diversity		
a) Relative area by forest type	8	8
b) **Forest type area by structure**	(9)	4
c) Forest type area protected	3	7
d)		

Stands change from one structure to another at different rates based on growth (dashed arrows) and disturbance regimes (solid arrows) based on aspect, species, etc.

Savanna

Open

Complex

Dense

Understory

Fig. 6.3. Example of the type of information used to develop scores for *Indicator 1.1b, Country B* (Table 6.2), forest type area by structure. Different forest types would be expected to contain all or some of these structures, or different classifications of structures specific to the forest type.

Country A's Criterion 1.1b indicator means that 18 of the 20 forest types contain the minimum required area in each structure needed by native species - and that two forest types do not contain enough of all structures (Table 6.4).

Information for each forest type would come from individual administrative regions, forest management districts, or other logical geographic units, providing the information *upward* to each forest type module. The information could include actual area, potential area, and sub-area of each structure - with the measurable classification of forest structures for each forest type being *handed down* to the regional modules from the level of the forest type module. Within each region,

Table 6.4. The score for *Indicator 1.1b (forest type by structure), Country B* (Figure 6.3) would be based on measurable characteristics and flow upwards through the hierarchy of sustainability criteria.

Forest Types	Area (Mha)	Savanna Current	Savanna Minimum needed?	Open Current	Open Minimum needed?	Dense Current	Dense Minimum needed?	Understorey Current	Understorey Minimum needed?	Complex Current	Complex Minimum needed?	Number of targets achieved
EASTERN REGION												
White-Red-Jack Pine	271	30	27	28	27	111	27	54	27	49	27	5
Spruce-Fir	1483	15	0	151	150	747	150	300	150	270	150	5
Longleaf-Slash Pine	124	21	20	13	12	44	12	24	12	22	6	5
Loblolly-Shortleaf Pine	142	8	14	14	14	66	14	28	14	25	14	4
Oak-Pine	180	20	18	19	18	73	18	36	18	32	18	5
Oak-Hickory	1970	110	197	106	197	1375	197	200	197	180	197	3
Oak-Gum-Cypress	1012	10	100	101	100	521	100	200	100	180	100	5
Elm-Ash-Cotton Wood	404	4	40	40	40	208	40	80	40	72	40	5
Maple-Beech-Birch	1014	10	100	101	100	524	100	200	100	180	100	5
Aspen-Birch	481	5	40	50	48	296	48	100	48	30	48	3

Forest Types	Area (Mha) Current	Savanna Current	Savanna Minimum needed[1]	Open Current	Open Minimum needed[1]	Dense Current	Dense Minimum needed[1]	Understorey Current	Understorey Minimum needed[1]	Complex Current	Complex Minimum needed[1]	Number of targets achieved
WESTERN REGION												
Douglas-Fir	3 080	180	300	309	300	1 451	300	600	300	540	300	5
Ponderosa Pine	2 774	305	277	292	277	1 125	277	554	277	499	277	5
Western White Pine	49	6	5	5	5	19	5	10	5	9	5	5
Fir-Spruce	4 720	47	470	472	470	2 415	470	940	470	846	470	5
Hemlock-Sitka Spruce	423	4	420	42	42	217	42	84	42	76	42	5
Larch	155	9	15	15	15	74	15	30	15	27	15	5
Lodgepole Pine	3 321	183	330	339	330	1 545	330	660	330	594	330	5
Redwood	50	1	5	3	3	35	3	6	3	5	3	4
Other Western												
Softwoods	1 595	16	160	161	160	810	160	320	160	288	160	5
Western Hardwoods	1 945	20	200	201	200	964	200	400	200	360	200	5

18 out of 20 forest types are acceptable. Normalized rank = 9.

¹ Minimum area needed is assumed to be 10% of the total remaining area of that forest type.

information would be amalgamated by land ownership, geographic area, or other sub-groupings, until each stand is appropriately classified.

Other criteria and indicators reflecting management objectives and resource values could similarly be measured, grouped, and *handed up* in the same way. The appropriate methodology for grouping would be developed by each group or module at a certain level and handed down for use by the next lower hierarchical level, to ensure these different groups are handing up consistently measured information.

It is unclear if, and how, the various indicators of biological diversity (shown in Table 6.2, for example) could be amalgamated into a single overall criterion of biodiversity. If, however, such synthetic indicators could be developed for each of the seven criteria identified in Table 6.1, then this could assist policymakers by reducing the complexity, and therefore some of the delays and mistakes in their decision-making.

Considerable work is needed in the development and definition of the appropriate measures in order that the indicators be robust. The indicators ought to be applicable across widely varying forest management conditions, be reliable in on-the-ground measurement and reporting, and be capable of validating significant aspects of sustainability which are generally agreed to be important (see Section 4 below and Chapter 7 in this volume). More effort is needed in determining both the measures themselves, and the most effective ways of communicating the appropriate amount of information across hierarchical levels.

3.2 Applications of Sustainability Criteria to Forest Management Decision-Making

The systems approach illustrated in the above example demonstrates the use of measurable criteria in general assessment of sustainable forestry at both the international and the increasingly more localized hierarchical levels. Measurable criteria, though, are also useful for determining the effects of different management alternatives on various objectives, with slight modifications as shown in the following example. When choosing among different management strategies, it may be considered desirable for biodiversity to increase over time. The inevitable change in stand structures over time across a landscape can be calculated using existing computer programs that automate the many calculations needed for integrating the measurable criteria to higher spatial scales. A suite of such forest models can be envisaged to link hierarchically across scales, ranging from the spatially explicit stand models needed to examine criteria and indicators for complex stands, through local landscape ecosystem management simulators.

One such program, Landscape Management Systems or LMS (McCarter *et al.*, 1998), can integrate information from an individual tree level (*e.g.* height, diameter, species, crown lengths) through the stand level (*e.g.* stand structure class, trees per acre, volumes per acre by log sorts) to the landscape level (percent of area in each structure and wind, fire, insect hazard class, log volume, or financial

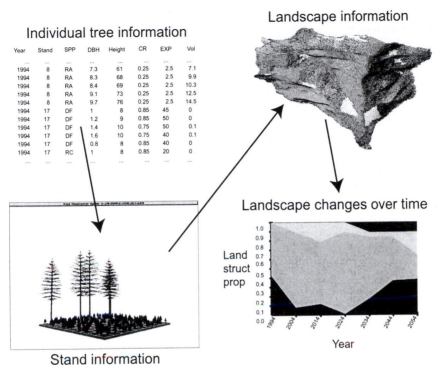

Fig. 6.4. Information can flow *upward* in the hierarchy from the individual tree level and be amalgamated and displayed in tabular, graphical, and/or visualization forms, as well as over time. Computer programs are automating some of the routine calculations required for processing this information (McCarter *et al.*, 1998).

concern flows). Output information can be expressed in various tables, charts, and visualizations (Figure 6.4). Different landscape areas can be combined within a forest type to ensure values are achieved at broader levels (Oliver *et al.*, in press); however, a simple indicator is required that will show the relative merit of each change in values. For example, assume *Country B* (Table 6.2) faced three policy alternatives that could change the value of forest types over the 80-year planning period as demonstrated in Figure 6.5. The challenge is to express these alternatives in simple, measurable criteria so that policymakers can readily understand the differences and merits of each. All three alternatives have an average score of 6 over the 80-year period; although one alternative shows a slow increase to *7*, followed by a stable score; another shows a rapid increase to *10*, and then a decline to *3*; the third shows no increase for several decades, followed by a jump and maintenance of *10* for the final two decades. The application of visualization techniques may assist policymakers in determining the differences between these alternatives in a quick, easily understandable fashion (see *Part V* of this volume).

| | Decade | | | | | | | | |
	1999-2009	2009-2019	2019-2029	2029-2039	2039-2049	2049-2059	2059-2069	2069-2079	Average
Management #1	4	5	6	6	7	7	7	7	6
policy #2	4	7	9	10	8	6	4	3	6
alternative #3	4	4	4	4	4	9	10	10	6

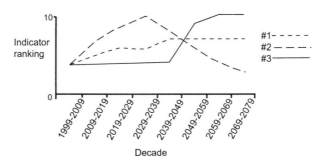

Decadal change in ranking of indicator
1b with three alternatives.

Fig. 6.5. Robust techniques are needed for showing subtle differences in how an indicator would change with different policy-management alternatives. Three example alternatives for future management of forests of *Country B* (Fig. 6.3) *Indicator 1.1b*; all have same average value, but dramatically different effects over time.

Once an alternative management scenario has been decided upon at the international, national, or regional/administrative unit level, the scenario, as well as the expected output measures, can be *disaggregated* to each more area-specific group. These expected output measures then become the target objectives of management - and the things to monitor to ensure that the objectives are met.

4   Outlining Suitable Criteria: The Montreal Process

Historically, deforestation and subsequent erosion have led to timber production and water protection being the overriding objectives of forest management (Boyce and Oliver, 1996). As countries become economically more developed and global trading infrastructures are formed, timber and water quality shortages become less immediate (Oliver, in press); consequently, additional values and underlying societal demands emerge. Such values can be added to, or their relative weight prioritized within, pre-existing frameworks of criteria and indicators as part of the hierarchical systems approach described above, without the need for re-inventing a whole new system.

One of the principal processes by which new values for sustainable forest management have been recognized has been through major international initiatives on forest management (see Chapter 7 in this volume for discussion of a range of such initiatives). This section focuses upon the Montreal Process, which began when Canada convened an *International Seminar of Experts on Sustainable Development of Boreal and Temperate Forests* after the United Nations Conference on Environment and Development, Rio de Janeiro, June 1992. The Process' purpose has been to establish *Criteria and Indicators for the Conservation and Sustainable Management of Temperate and Boreal Forests*. Whether or not these criteria and indicators are adopted, modified, or replaced by others, they provide a well thought-out framework with which to advance discussion of forest management and to begin constructing hierarchical systems for sustainable forest management. These values (criteria) and summaries of some of their more specific measures (indicators) are discussed and interpreted below.

4.1 Criterion 1: Biological Diversity

Biological diversity is typically viewed from two perspectives: (1) protecting habitats for all native species by ensuring all forest types and stand structures within forest types exist and are distributed throughout the world; and (2) protecting individual species known to be in danger of extinction. It is important also to consider temporal diversity (*e.g.* patterns and representation of different seral stages), as well as the genetic diversity of the local species.

With the realm of economic development, agriculture is often concentrated on productive soils, while less productive soils are allowed to return to forested conditions; the result is a net increase in forest area, but a decline in area of forest types occupying productive soils. Additionally, the regrowth of forests, efficient fire protection, reduction of fuelwood use, and movement of people to cities has led to many forests growing to what has been characterized as a *dense* structure (Oliver, in press; see Figure 6.3). A further result of these trends has been an increase in the number of endangered species dependent upon forest openings, savannahs, and complex structures (Young, 1992; Oliver *et al.*, 1997).

The following could be combined as measures of these indicators:

- current, projected, and potential area by forest types;
- current and projected area in each structure;
- status of endangered species; and
- current and projected areas in reserves.

4.2 Criterion 2: Productive Capacity

The ability of forests to provide future timber and other commodities may be viewed in three ways: (1) the forest's area, (2) the forest's timber volume, and (3) the forest's timber growth/removal ratio. While the world's forest area has declined from 8 billion hectares to 3.45 billion hectares during the past four thousand years (FAO, 1993), it ought to be recognized that countries that have undergone economic development have actually increased their forest area, their standing forest volume, and their timber consumption during the past century (FAO, 1997). This increase may be due to four trends in forest management practice: (1) the conversion of marginal farming and grazing land forests; (2) longer rotations due to a curtailment of wood harvested for sustenance; (3) the control of fire and the implementation of modern silviculture methods; and (4) a reduced need for draft animals (and pastures) due to an increase in the use of automobiles and tractors. Exceptions occur where cessation of harvest in one region for environmental reasons has led to excessive harvests in other regions until global markets have adjusted or until a disturbance regime, such as fire, affects a growing forest (Oliver *et al.*, 1997; Perez-Garcia *et al.*, 1997). Less developed countries are still experiencing a net reduction in forest area and volume; this reduction has been associated with subsistence agriculture and extreme use of wood for fuel.

The following could be combined into measures of these indicators:

* current and projected forest area;
* timber volume;
* tree growth; and
* harvest rates by species, sizes, and forest types.

Increasingly, it is important in both affluent and less developed countries also to measure the productive capacity of the forest for many non-timber products, such as mushrooms, medicinal plants, fish, and wildlife. This strongly suggests that multiple measures of productivity may be required in evaluating sustainability, rather than an exclusive focus on timber productive capacity. It is important to include broader measures describing the status of overall site productivity, and consider the interaction between disturbance, seral stage sequences, and long term site productivity.

4.3 Criterion 3: Forest Health and Vitality

An increase in global transportation has contributed to the spread of exotic insects, diseases, and other plant and animal pests to many places in the world. The increase of exotic pests in reserved areas (*e.g.* The Great Smokey Mountain National Park, USA; Campbell and Schlarbaum, 1994) suggests that proactive management may be more effective in combating these pests than attempting to exclude them by prohibiting active management.

Air pollution provides an example of an area-specific problem that has been solved primarily by non-forest activities, such as factory and automobile pollution control. If overly crowded forests in developed and developing countries are allowed to be devastated instead of being properly managed, air pollution from forest fires will continue to increase. Current and projected spatial extents and effects of each forest pest, pollution problem, and other disturbance patterns associated with the imbalances of forest conditions could be combined into measurements for these indicators.

4.4 Criterion 4: Soil and Water Conservation

Poor water quality, soil erosion, and soil compaction are generally associated with uninformed agriculture and grazing practices by people engaging in subsistence farming and ranching. These adverse conditions can be caused by poor forestry practices, and also by unnaturally hot fires which burn in unmanaged, overly crowded growing forests in more developed countries; however, as countries become developed, such causes of these adverse conditions decline. Inventories and projected areas of erosion, potential erosion, and high risks of fire, insect outbreak, windthrow, and other catastrophic events that can lead to erosion, loss of soil fertility, and water quality problems could be combined with inventories of aquatic and riparian systems to provide measurements for these indicators.

4.5 Criterion 5: Contribution to Global Carbon Sequestration

Carbon dioxide is taken out of the atmosphere by growing forests and is added to the atmosphere by burning or rotting trees (or forest products). Carbon dioxide is kept out of the atmosphere if wood is used in construction instead of substitute products such as steel, aluminium, brick, and/or concrete; in fact, these substitute products require more fossil fuels for their refinement and production than wood products do. Carbon dioxide can also be kept out of the atmosphere by using wood as an energy source instead of fossil fuels.

Analyses of the relative amounts of carbon dioxide removed/prevented by forest growth, wood use for substitute construction products, and wood use for energy have shown that wood use for substitute products contributes the most to carbon sequestration (Kershaw *et al.*, 1993; Perez-Garcia *et al.*, 1997). An effective means of managing forests for their contribution to global carbon sequestration would be to institute a three pronged management process that would: (1) grow forests on relatively long rotations to take carbon dioxide out of the atmosphere; (2) harvest and utilize wood for high quality construction products that avoid the use of non-wood substitutes; and (3) utilize the residuals for wood energy. Inventories and projections of forest volumes, forest fire risk, and timber harvest could be combined to provide measurements for these indicators.

4.6 Criterion 6: Socio-Economic Benefits

Along with economic development, the abandonment of marginal farming and grazing lands, and the mechanization of farming, there has been a migration of people to cities and a concomitant increase in urban problems. Advances in communication and transportation are enabling people to live in forested, rural areas while maintaining the high quality life styles traditionally associated with urban environments. Currently there are concerns that forested areas may not have the infrastructure of roads, safe forests (from fires and hostile animals), the diversity of habitats, a skilled workforce, adequate equipment, or other amenities to accommodate a reverse migration (Oliver, in press). This sort of infrastructure can be maintained by a diverse economy that effectively defrays the costs entailed in supporting any single resource sector. Current and projected employment can be estimated by various means; additionally, current and projected road concentration, recreation use and type, and other uses can be projected for each area and combined to provide measurements for these indicators. There is, however, a particular need to develop more robust criteria for assessing spiritual, aesthetic, and other cultural and quality-of-life values which reflect public concern over their environment, their community identity, and human health (as discussed in more depth in Chapter 7 in this volume).

4.7 Criterion 7: Legal, Institutional, and Economic Frameworks

The legal, institutional, economic and research frameworks are actually the policy and management tools that can be used to achieve the other criteria. Many of the other criteria represent values that have not traditionally been incorporated into forest management (or even into society in general). Consequently, attempts to mandate these values without the development of innovative methods to achieve them can readily create inequities, such as the expectation that private individuals will provide public values at their personal expense (Lippke and Oliver, 1993); expectations such as these are rarely fulfilled. For example, a mandate that forests containing spotted owls in the Pacific Northwestern United States could not be harvested has actually decreased spotted owl habitat. Private landowners rapidly harvested their old forests before spotted owls could nest in them, for fear of being prevented from harvesting in the future and being denied the resulting revenue.

Various economic incentives, legal baselines, education techniques, and research programmes can change management behaviour, economic advantage, and perhaps even human perception (as discussed by Daniel and others in this volume), in order to allow forests to provide the desired values. However, the limitations of management systems such as forest tenures in Canada (which some would argue effectively preclude ecosystem management on public lands due to the requirement for timber management with ecological constraints), need to be taken into account in any evaluation of sustainability. Economic analyses (*e.g.* present net worth, return on investment, return as a percent of inventory, and cash flow) and sensitivity analyses, in addition to the survey of different laws, regulations, and data on behaviours of landowners, can be combined to provide measurements for these indicators.

5 Conclusions

A case has been presented for a systems approach to evaluating sustainable forest management and supporting management decisions, using a hierarchical framework to relate these measures across geographic scales or management/ policy levels, from the stand to the globe. This approach can be used to characterize and analyse such complex interactions as temporal changes in stand structure or seral stage distributions, through means such as simple scores anchored to relevant indices or thresholds,

Each of the criteria and indicators described in Section 4 above can be expressed at each of the hierarchical levels in a country or management agency. In this way, the reality of variability at small spatial scales is incorporated up to national-level criteria and indicators, while retaining the possibility of disaggregating down to real ecosystems. Much work is needed to develop robust ways to measure and display each criterion and indicator, as well as their changes over time, at each hierarchical level. Work is also needed to develop better ways to obtain and summarize the information used to calculate compliance with the criteria. Ways to develop creative management/policy alternatives and to ensure adequate implementation and monitoring are also needed. These needs must be met in ways that are simple enough to avoid complicated bottlenecks at any level of the management hierarchy, yet are robust enough to allow the desired objectives to be achieved.

Clearly, there are many hurdles to overcome in developing a generally agreed set of robust criteria and indicators for sustainable forest management. The simpler examples such as those provided in this chapter can be criticized perhaps for being arbitrary or overly simplistic: one cannot easily apply global averages of original forest area to local circumstances, for example; and the threshold chosen for any "yes/no" standard of sustainability, *e.g.* minimum area of forest type needed, should be supported by solid research findings if it is to be defended and not warped by either the forest industry or an overly zealous public interest lobby. It is also vital that not all forest analysis be reduced to very simple measures aimed at high-level politicians and decision-makers: the complexity and variability in the ecosystem must be communicated.

Nevertheless, a key component in meeting the objectives of sustainable forest management is the adoption of meaningful and measurable criteria. If the current and desired condition of the forest can be expressed in robust and simple measures, it will then be possible to develop a process to adjust management to achieve different objectives without the confusion and rancour currently occurring in much of the world's forestry arena. Similar measurable criteria (*e.g.* stock market indexes, prime interest rates, profit/earnings ratios) have assisted the global financial community operate with a greater level of stability while allowing dramatic change.

References

Botkin, D.B., and M.T. Sobel. (1975) Stability in time-varying ecosystems. *The American Naturalist* 109 (970): 625-646.
Boyce, S.G., and C.D. Oliver (1996) The history of research in forest ecology and silviculture. In: H.K. Steen (ed) *The History of Forestry Research in the United States*. Forest History Society, Durham, North Carolina.
Brundtland, G.H. (1987) *Our Common Future*. World Commission on Environment and Development, United Nations, Oxford University Press, New York. 400 pp.
Campbell, F.T. and S.E. Schlarbaum (1994) *Fading Forests: North American Trees and Exotic Pests*. Natural Resources Defense Council Report. 47pp.
Canadian Council of Forest Ministers (1998) *National Forest Strategy, 1998-2003: Sustainable Forests, a Canadian Commitment*. Canadian Council of Forest Ministers, Ottawa.
Food and Agriculture Organization of the United Nations (1993) *Forest Resources Assessment 1990: Tropical Countries*. FAO Forestry Paper 112. FAO Forestry Department, United Nations, Rome.
Food and Agriculture Organization of the United Nations (1997) *State of the World's Forests*. United Nations Food and Agriculture Organization of the United Nations, Rome. <www.fao.org--forestry-publications>.
Kershaw, J.A., C.D. Oliver and T.M. Hinckley (1993) Effect of harvest of old growth Douglas-fir stands and subsequent management on carbon dioxide levels in the atmosphere. *Journal of Sustainable Forestry* 1:61-77.
Lippke, B., and C.D. Oliver (1993) How can management for wildlife habitat, biodiversity, and other values be most cost-effective? *Journal of Forestry* 91: 14-18.
Maalouf, A. (1984) *The Crusades Through Arab Eyes*. Random House Trade, New York.
McCarter, J.M., J.S. Wilson, P.J. Baker, J.L. Moffett, and C.D. Oliver (1998) Landscape management through integration of existing tools and emerging technologies. *Journal of Forestry* 96(6): 17-23.
Oliver, C.D. (1992) A landscape approach: achieving and maintaining biodiversity and economic productivity. *Journal of Forestry* 90: 20-25.
Oliver, C.D. (in press) The future of the forest management industry: Highly mechanized plantations and reserves or a knowledge-intensive integrated approach? Submitted to *Forestry Chronicle*.
Oliver, C.D. and B.C. Larson (1996) *Forest Stand Dynamics*. Update edition. John Wiley and Sons, New York. 521 pp.
Oliver, C.D. and M. Twery (in press) Decision support systems: models and

analyses. Paper to be published in *Proceedings of the Ecological Stewardship Workshop, Tucson, Arizona*. USDA Forest Service, December 1995, 60 pp.

Oliver, C., D. Adams, T. Bonnicksen, J. Bowyer, F. Cubbage, N. Sampson, S. Schlarbaum, R. Whaley, H. Wiant and J. Sebelius (1997) *Report on Forest Health of the United States by the Forest Health Science Panel*. A panel chartered by Charles Taylor, Member, United States Congress, 11th District, North Carolina. Summary: 72 pp. Main document: 334 pp.

Oliver, C.D., M. Boydak, G. Segura and B.B. Bare (in press) Forest organization, management, and policy. In: M.L. Hunter, Jr. (ed) *Maintaining Biodiversity in Forest Ecosystems*. Oxford University Press.

Perez-Garcia, J.P., C.D. Oliver and B.R. Lippke (1997) *How Forests Help Reduce Carbon Dioxide Emissions to the Atmosphere*. Report to the Subcommittee on Forests and Forest Health of the Committee on Resources, United States House of Representatives, July 7, 1997.

Sprugel, D.G. (1991) Disturbance, equilibrium, and environmental variability: What is "natural" vegetation in a changing environment? *Biological Conservation* 58: 1-18.

Young, M.R. (1992) Conserving insect communities in mixed woodlands. In: M.G.R. Cannell, D.C. Malcolm, and P.A. Robertson (eds) *The Ecology of Mixed-Species Stands of Trees*. Blackwood Scientific Publications, London. pp. 277-296.

Chapter seven:

International Initiatives for the Sustainable Management of Forests

Jeffery Burley
Director, Oxford Forestry Institute and President, International Union of Forestry Research Organizations
Oxford

1   Introduction

The International Union of Forestry Research Organizations (IUFRO) is the oldest international, non-governmental research organization in the world, having been created in 1892, and it now comprises some 700 Member Institutions and 76 Associate Members in 109 countries with 276 collaborative working units (Divisions, Research Groups, Working Parties, Task Forces and a Special Programme for Developing Countries). With approximately 15,000 scientists these units cover virtually all topics related to forests and forestry. The Task Force on Sustainable Forest Management is of particular relevance to the subject of this workshop.

Recently IUFRO has sought to achieve a greater voice in international policy processes, to enhance inter-disciplinary research, and to prepare (by the time of the 21st IUFRO World Congress in Malaysia during August 2000) state-of-knowledge reports that will summarize what is known about the major policy-related issues and what is not known. In particular we have sought to encourage researchers to direct their efforts to the solution of policy questions. For too long researchers have requested their political and administrative masters to fund research but have failed to demonstrate adequately the value of the information so far obtained or the genuine need for new research.

In a book based on a meeting in the University of British Columbia we may expect a concentration on the issues and practices of BC and Canadian or North American forests and foresters; my role here is to draw attention to the fact that there are other forests and that considerable efforts are being made worldwide to develop practices of sustainable forest management. The world's total forest and woodland estate approximates 4 billion hectares and these are split almost equally between tropical moist, tropical dry and temperate forests. The tropical moist forests of west and central Africa, Asia and Latin America contain the land's richest ecosystems in terms of biological diversity and biomass and provide 15% of the world's commercial timber plus the livelihoods of 140 million people. The tropical dry forests and woodlands are largely located in Africa while the temperate and boreal forests are principally in industrialized countries.

The benefits (products and services) obtained from forests and trees vary in relative importance between regions and local sites but basically comprise three groups: economic products (timber, reconstituted wood, chemical derivatives, energy, human food and animal forage or fodder); environmental services (including soil conservation, water supply and flood control, climate amelioration, site rehabilitation, and genetic conservation); and social benefits (including employment and income generation throughout the year, risk reduction, diet diversity, human and animal health, education, community empowerment, recreation and cultural needs). With such multiple benefits possible and often demanded, it is commonly difficult to manage the forest in ways that can optimize their provision while minimizing conflicts. However, sustainable development in general requires the balanced use of human skill, natural resources and finances and thus sustainable forest management itself must seek such balances.

2   International Sustainability Agreements

There have been hundreds of definitions of sustainability and sustainable forest management but a commonly accepted concept is that of the Inter-Ministerial Conference on European Forests in Helsinki, Finland, in 1993 - the *Helsinki Process* - Sustainable management means the stewardship and use of forest land in a way, and at a rate, that maintains their biodiversity, productivity, regeneration capacity, vitality and their potential to fulfil now, and in the future, relevant ecological, economic and social functions at local, national, and global levels; and that does not cause damage to other ecosystems (see Ministerial Council for the Protection of Forests in Europe, 1993).

Resulting from the United Nations Conference on Environment and Development in Rio de Janeiro, Brazil, during 1992, a number of products, initiatives and processes were established that were relevant to sustainable forest management. Immediate results included the Agenda 21 itself and the Conventions on Biological Diversity, Climate Change, and Desertification (joining the existing Convention on International Trade in Endangered Species). The International Tropical Timber Organization produced the first set of principles for tropical forest management, and subsequent inter-governmental initiatives included the Helsinki Process (for European forests), the Montreal Process (for boreal and temperate forests outside Europe), the Tarapoto Proposal (Amazon forests), African Timber Organization (African moist forests), Lepaterique Process (Central American forests), and various conferences of international organizations (UNEP/FAO, FAO/

UNEP and FAO/ITTO) for dry Africa and the Near East plus the harmonization of the various sets of criteria and indicators. The origins and contents of all these initiatives, processes and their criteria and indicators, were described by Grayson and Maynard (1997) and summarized in FAO (1997a); for a description of the formulation of sustainable forest management standards see Lammerts van Bueren and Blom (1997).

Principles and criteria seek to identify the critical components of sustainable forest management and to operationalize sustainability, providing endpoints and a recognizable identity. Indicators are the primary tools of evaluation and may be generic or site specific with ranking possible according to importance. Verifiers are extremely site specific and may have defined thresholds.

Despite the variability of these initiatives and their relevant forests, there has been a convergence on the criteria for sustainable forest management; indeed it might be argued that any group of university forestry students locked in a room for a day would arrive at the same list: biodiversity (landscape diversity, species diversity and genetic variation); productivity (of the ecosystem); soil conservation (including erosion and natural hazards); water conservation (including water quantity and water quality); forest ecosystem health and vitality (ecosystem functioning); contribution to global ecological cycles; and ability of the forest ecosystem to fulfil socio-economic functions.

However, the indicators by which these criteria might be evaluated are much more debatable and for some criteria no good quantitative indicators have yet been developed, particularly for social benefits; hence the value of this present workshop which addresses aesthetic issues that have not been considered seriously in any of the international processes. The problem of sampling appears in all indicators; for example, the assessment of biodiversity varies from intra-specific to inter-specific and ecosystem levels, and different species may vary in these estimates at different sites, ages of host trees, or season, etc. Similarly in sampling human preferences, estimates of age, gender, economic, religious, cultural and educational backgrounds must be taken into account.

3   Amenity and Aesthetic Indicators of Sustainability

All of the major international initiatives include a criterion related to socio-economic functions, essentially stated as the maintenance and enhancement of national and local multiple socio-economic benefits. Typically these include some or all of the following: share of the forest sector from the gross national product; changes in the rate of employment in forestry; provision of recreation; cultural, social and spiritual needs and values; participation among stakeholders; reduction in the number of forest offences.

Technically, aesthetics and amenity are not the same thing, but the terms tend to be used more or less synonymously. *The Oxford English Dictionary* defines aesthetic as "belonging to the appreciation of the beautiful; having such appreciation; in accordance with principles of good taste"; it derives from the Greek *to perceive*. The dictionary defines amenity as "pleasantness (of places, persons, etc.); pleasant ways"; this derives from the Latin *to love*. A modern application of *amenity* is the provision of facilities (*e.g.* a campsite, running water, telephone etc.). However, for the purposes of this debate, the two terms are treated synonymously and include both the pleasing appearance of forests and the recreational or cultural use of them.

Three of the forest benefits listed above are partially related to amenity and some indicators have been developed. Indicators of the provision of recreation include the area of forest with access per inhabitant; the percentage of total forest area; the number of facilities available for recreation; and the number of visitor-days attributed to recreation. Indicators of cultural, social and spiritual needs and values include the area and percentage of forest managed for cultural needs; the non-consumptive uses of forest; and the area and percentage of forest used for supporting local populations. Participation among stakeholders is indicated by the level of local population participation in the management of forest activities; the quality of life of local populations; and the rate of improvement in the livelihoods of forest-dependent communities.

In many countries, at the national level, social use indicators include: numbers of forest employees; extent of employment multiplier effects; rates of afforestation in areas of multiple forest use; numbers of forest visitors; existence of monitoring of planting in sensitive areas; existence and efficiency of urban/peri-urban woodland inventories; existence of national opinion surveys, public awareness and public involvement in forest conservation. Comparisons of such indicators in four international processes and several developing countries were provided in a valuable working document by FAO (1997b).

At the level of the forest management unit, the following indicators are assessed: extent of access available on public rights of way; extent and precision of information about recreational facilities and access communicated to the public; number of opportunities for recreational pursuits in woodlands; the impacts of recreational activities on other forest users; the extent of efforts to mitigate vandalism and anti-social behaviour; the consideration of requests to use woodland for environmental education; and the extent of consultation and involvement of local communities. It is clear that many of these possible indicators cannot be easily quantified other than by social survey techniques and personal judgements.

Based on such criteria and indicators some countries have developed forestry standards that are now accepted by governments and non-governmental organizations. In the United Kingdom (UK) the Forestry Standard (Forestry Authority, 1998) includes eight major criteria: forest soil condition; water quality, yield and discharge patterns; net carbon sequestration and air pollution; timber and wood production; nature conservation in and around forests; forest workforce competence and safety; social use and participation in sustainable forest management; and historic, cultural and aesthetic use of forests (Her Majesty's Government, 1994; Forest Stewardship Council, 1998; UK Woodland Assurance Scheme, 1999).

Within the UK Forestry Standard, at the national level, specific criteria for historic, cultural and aesthetic uses include: the conduct of surveys and registers of ancient monuments; reporting on damage to ancient monuments; the determination of woodland aspects of countryside character and landscape assessments; and the completion of survey reports for special areas such as National Parks. At the level of the forest management unit (typically a managed forest stand), UK indicators include the following: important sites are clearly recorded; sound principles for integrating archaeological sites in woodland are adopted; archaeological sites are protected and damage avoided; landscape principles of forest design are used; and the cultural and historical characters of a site are considered.

The methods for assessing social and aesthetic values are varied and they are variously applicable to specific sites or conditions; they include both qualitative methods, some of which may include quantitative elements (*e.g.* focus groups; interviews of individuals in person, by telephone or by questionnaire correspondence) and formal quantitative methods (including contingent valuation; willingness to pay or willingness to accept; expression of preference - particularly by visual imaging including photographs, videos and computer graphics, all with or without comparisons). In social surveys conducted in the UK and USA, differences have been recognized between static, visual perception of landscapes from a distance (with a sense of beauty, coherence, complexity and mystery) and engaging aesthetics (with sensory stimuli, psychological reactions of peace, mental rejuvenation and spiritual well-being) - see Chapter 10 in this volume.

Such detailed applications have not yet progressed far in developing countries. However, the Centre for International Forestry Research (CIFOR, Bogor, Indonesia) undertook during the 1990s various comparisons of some 1,400 criteria and indicators in a range of developed and developing countries. As *best bets* for social criteria in developing countries, CIFOR identified the following: local management is effective in controlling maintenance of and access to the resource; forest actors have a reasonable share in the economic benefits derived from forest use; and people link their own and their children's future with the management of forest resources (see Prabhu *et al.*, 1996; 1998).

The *best bet* indicators of social use defined by CIFOR included the following: clear ownership and rights to use of resources; rules and norms of research use are monitored and enforced; there is a means of conflict resolution; access to resources is perceived as fair and secure; equitable benefit sharing; employment and training opportunities exist; wages are fair and damages are compensated in a fair manner; local investment in surroundings; out-migration levels are low; population number balanced with natural resource use; children educated about natural resource management; and indigenous people maintain spiritual links to the land and their familiar landscape features (this latter issue is of particular importance in British Columbia - see Chapter 8 in this volume).

If the CIFOR indicators are compared with those of the UK Forestry Standard, all are similar except for the lack of recreational and aesthetic uses in developing countries. This may reflect the overriding demand for products by local forest-dependent people but it ignores the increasing demand for recreation by indigenous people as education and affluence increase and also the demands of international tourists for ecological and recreational tourism with its consequent income generation for the host country.

To refine amenity indicators in such situations is particularly difficult but necessary. Within the UK, further development of social indicators derived from social surveys could include: human health (tree-related therapy, emotional well-being, physical exercise, and relaxed environment); education and information (nature trails, information boards, conservation volunteer systems, and links to schools and community groups); personal freedom and security (access to woodlands and freedom from fear or sense of threat); and culture, leisure and recreation (including relaxation, walking, climbing, cycling, horse-riding, wildlife viewing, camping, children's play, access to ancient monuments and sacred sites).

4 Conclusion

The challenges for the future thus include the development of appropriate methods for evaluating social aspects of sustainable forest management in general, the quantification of a range of indicator variables summarized above, and assistance to individuals and communities to recognize that, while beauty may be in the eye of the beholder, aesthetics is more than beauty. As defined earlier, aesthetics is often considered synonymous with amenity and thus implies both the appearance and the existence for use of the resource. In a sustainable landscape or for sustainable forest management in general, the balance between multiple competing objectives is compounded by the competing democratic demands of different users.

Criteria and indicators were developed principally to facilitate the monitoring over time of the sustainability of forest management in a particular forest or nation. This contrasts with the standards established by certifying agencies such as the Forest Stewardship Council (FSC) or Canadian Standards Association which are used to ensure comparable sustainability of management in different forests at one time. The former is a means of monitoring performance, the latter a means of defining equal performance. In practice there is convergence because some certification schemes involve a commitment to continuous improvement towards internally set targets, and hence monitoring performance is crucial. For both, the need for objective and quantifiable indicators is urgent. Again within my own country, the UK Government's Forestry Standard sets out criteria and indicators by which the Government will monitor sustainability at a national and forest management unit level. These have been combined with the FSC standards to allow the preparation of the UK Woodland Assurance Scheme Certification Standard (UKWAS, 1999). This required lengthy negotiation, to ensure the requirements of the standard were realistic, scientifically justified and still met FSC Principles and Criteria. Agreement on this was finally reached in mid-1999, and it is still the only

Certification Standard in the world which has the full support of all stakeholders - FSC, Government, NGOs, and forest owners, both State and private.

The 1999 Peter Wall Institute for Advanced Studies (PWIAS) workshop demonstrated excellently the need for, and outcomes of, inter-disciplinary co-operation - something that is difficult for many university institutions that are judged by their academic output of pure, mono-disciplinary science - and I would congratulate the PWIAS on its commitment to multi-disciplinary approaches to major issues such as including aesthetics in criteria for sustainable forest management.

Acknowledgement

I wish to thank my colleague Dr. Joanne Chamberlain (now at the Centre for Natural Resources and Development, Green College, Oxford) for her help in assembling material for this chapter.

References

FAO (1997a) *Forestry Information Notes - Criteria and Indicators of Sustainable Forest Management*. Food and Agriculture Organization of the United Nations, Rome.

FAO (1997b) *Integrating Criteria and Indicators of Sustainable Forest Management in the National Forest Programmes*. (Based on the work of S.O. Thirong Patrick, consultant.) Working document, Food and Agriculture Organization of the United Nations, Rome.

Forestry Authority (1998) *The UK Forestry Standard - The Government's Approach to Sustainable Forestry*. Prepared by the Forestry Commission and Department of Agriculture for Northern Ireland. The Forestry Commission, Edinburgh, UK.

Forest Stewardship Council (1998) *National Standards for Great British Forest Management*. Forest Stewardship Council, Llandiloes, Wales, UK. <www.fsc-uk.demon.co.uk>.

Grayson, A.J. and W.B. Maynard (eds) (1997) *The World's Forests - Rio+5: International Initiatives Towards Sustainable Management*. Commonwealth Forestry Association, Oxford.

Her Majesty's Government (1994) *Sustainable Forestry: the UK Programme*. Her Majesty's Stationery Office, Command Paper CM 2429, London.

Lammerts van Bueren, E.M. and E.M. Blom (1997) *Hierarchical Framework for the Formulation of Sustainable Forest Management Standards*. The Tropenbos Foundation, Leiden.

Ministerial Council for the Protection of Forests in Europe (1993) *Resolution H1. General Guidelines for the Sustainable Management of Forests in Europe.*

Ministerial Council for the Protection of Forests in Europe, 16-17 June, 1993, Helsinki.

Prabhu, R., C.J.P. Colfer, P. Ventakeswarlu, L.C. Tan, R. Soekmadi and E. Wollenberg (1996) *Testing Criteria and Indicators for the Sustainable Management of Forests: Phase 1, Final Report.* CIFOR Publication, Centre for International Forestry Research, Bogor, Indonesia.

Prabhu, R., H.J. Ruitenbeck, T.J.B. Boyle and C.J.P. Colfer (1998) Between voodoo science and adaptive management: the role and research needs for indicators of sustainable forest management. In: Proc. IUFRO/FAO/CIFOR International Conference on Indicators for Sustainable Forest Management, *Fostering Stakeholder Input to Advance Development of Scientifically-Based Indicators*, 24-28 August, 1998, Melbourne.

UKWAS (1999) *Certification Standards. Woodland Assurance Scheme.* Forestry Commission, Edinburgh, <http://www.forestry.gov.uk>.

Chapter eight:
The Tloo-qua-nah Principle in Forest Sustainability: A First Nations Perspective

Umeek (E. Richard Atleo)
First Nations Studies, Malaspina University-College
Nanaimo, British Columbia

1   Introduction

The purpose of this chapter is to present some illustrations of an aboriginal *world view* and behaviour that relate to sustainable uses of forest resources; specifically, I address how the holistic perspective on all living things (biodiversity) differs from typical Western, or industrial attitudes toward forestry and the importance of balance and respect in human relationships with the land.

One literal translation of *Tloo-qua-nah* is "we remember reality". In precontact times among the Nuu-chah-nulth it was a sacred ritual which re-enacted the constant struggle between creative and destructive forces among life forms. In the Tloo-qua-nah, the destructive forces are enacted by an abduction of people by wolves that represent a destructive loss to the community. The abducted people are ultimately rescued or *saved* and, in the process of being restored back to the community, become better people as a result. The abduction is recognized as an example of what happens when humans become slack in vigilance and discipline because the dangers of destructive loss are ever present. The Tloo-qua-nah then becomes a forum for *Ha-huupa*, or teachings, that emphasize how people should live.

The context for these teachings is found in the Nuu-chah-nulth assumption about the nature of Reality that *everything is one*. Within this unity of existence, physical and spiritual forces are in opposition, and must be managed, balanced, and harmonized. These forces include human conflict as well as conflict among all other life forms. The outcome of this management is intended to enhance life. This Tloo-qua-nuh principle then, can also apply to forest practices where humans have made mistakes and therefore have suffered destructive losses. The intent of restoration of the losses to the forest, is that people remember how all life forms should be treated, with respect, as part of the unity of existence. From a First Nations perspective then, the question is whether *sustainability* can be rescued from the destructive forces of clearcut logging practices.

2   Tloo-qua-nah Principle in Sustainability

He-shook-ish t'sawalk means *everything is one* in the Nuu-chah-nulth language. In the beginning of creation, every life form was of one *thli-muhks-ti*, one spirit, one soul. Then, as told in the Nuu-chah-nulth story, a prophetic word came that someone was coming to change everything. Since no one likes change, preparation was made to resist. In self-defence one person made sharp knives from mussel shells but He who came to change everything transformed these knives into ears and that's how deer came to be. Whatever means that people employed to resist the impending change, became the instrument of that change and that is how biodiversity came to be. In outward form, everything changed, and there appears, today, to be a multiplicity of life forms, but the Tloo-qua-nah (to remember Reality) is a constant reminder of the origin of species. In the beginning everything was one, Heshook-ish t'sawalk. More than this, the Tloo-qua-nah is a constant reminder that life is essentially a struggle between creative and destructive forces that situates the *Quus* (the people) within, rather than above or beyond, the interconnected, interrelated, and interdependent web of life.

From origin stories, like the one recounted above, there developed a worldview that was articulated by Black Elk, a Lakota Sioux of the Pine Ridge Reservation in South Dakota, in the following way:

> We regard all created beings as sacred and important, for everything has a *wochangi*, or influence, which can be given to us, through which we may gain a little more understanding if we are attentive. We should understand well that all things are the works of the Great Spirit. We should know that He is within all things; the trees, the grasses, the rivers, the mountains and all four-legged animals, and the winged Peoples; and even more important, we should understand that He is also above all these things and peoples. (Brown, 1986, pp. 38-39)

In British Columbia, Canada, two hereditary chiefs of the Gitksan and Wet'suwet'en authored a book entitled *The Spirit in the Land*. Gisday Wa and Delgam Uukw (1992) explain the relationship between life forms as a law:

> The land, the plants, the animals and the people all have spirit - they all must be shown respect. That is the basis of our law. (p. 7)

Later in the same publication there is a comparative analysis between the Western world-view and the Gitksan and Wet'suwet'en world-view.

The Western world-view sees the essential and primary interactions as being those between human beings. To the Gitksan and Wet'suwet'en, human beings are part of an interacting continuum which includes animals and spirits. Animals and fish are viewed as members of societies which have intelligence and power, and can influence the course of events in terms of their relationship with human beings....The Gitksan and Wet'suwet'en believe that both humans and animals, when they die, have the potential to be reincarnated. But only if the spirit is treated with the appropriate respect. If bones of animals and fish are not treated with that respect, thereby preventing their reincarnation, then they will not return to give themselves up to humans (p. 23).

This interrelationship between life forms is the basis of Gitksan and Wet'suwet'en law as it is with Nuu-chah-nulth law. The Nuu-chah-nulth origin story explains the nature of the beginnings of these relationships. All life was of one spirit. In a sense all life is still one in spirit and that is the reason for continued recognition and practice of respectful protocols to ensure sustainability. The fact that some life forms are people and some are salmon and some are deer is a measure of the creativity of the divine. Since all life forms come from the same source, the Creator, all life forms are sacred and all are to be respected.

This is the reason for the honour and respect paid to the arrival of the first salmon of the season. If the salmon is not properly recognized and respected by established protocols then a mutually interdependent relationship is jeopardized. If diplomacy is discontinued beyond a certain tolerance point then the salmon may simply refuse to return. In modern terminology, over fishing, together with destruction and pollution of salmon rivers and streams, will eventually deplete stocks to the point of extinction. Many of the salmon streams in Clayoquot Sound "no longer support their historical numbers" (Bunnell and Atleo, 1995) and that is one consequence taught by the Tloo-qua-nah principle. Similarly, if a great tree is cut down without ceremony, if the tree is not recognized and respected, there are natural consequences not conducive to healthy relationships. Balance and harmony between all life forms was a standard, a goal of life that was not always achieved, but balance and harmony between life forms in the Americas was the norm prior to the arrival of the European. The Americas was seen as "pristine wilderness", by these new arrivals, rather than as a place of well managed ecosystems and highly developed protocols.

From a First Nations perspective, the reality of the spiritual realm is not an assumption but a collective lived experience. Stanley Sam (1992) in an unpublished manuscript entitled *Indian Doctor Stories* has said in this regard:

When our people sought spiritual power...they visited a very sacred pool...where they used special medicines.... Each family had its own special medicine which was handed down from generation to generation. Then he received a vision... from the animals. The animals who appeared in a vision were often otter, eagle, mink, squirrel, and a little brown bird.... Sometimes when these animals appeared in a vision they brought with them a certain rattle or song, or the special family-owned medicines.... [Dr. Atleo's] power was a land otter that turned into an eagle. He received this power in a vision.... The last Indian doctor of the Ahousaht was a woman. She became an Indian doctor by obtaining a vision of black chitons turning into snails. (pp. 1-4)

There is an element of egalitarianism in the practice of traditional Nuu-chah-nulth medicine. One went to a sacred place rather than to a central authority or institution. Although each family had their own way to acquire knowledge, power, and gifts, there were generalizable models of prayer, cleansing, fasting, meditation, and prayer songs as well as unmarked sacred areas outside of village boundaries. Spiritual power may be manifested in transformations, such as that of a land otter into an eagle or a black chitin into a snail. Year after year, generation after generation, indigenous people accessed the spiritual realm for information, power, gifts, and guidance in all matters.

If one received a gift of healing, it was a gift no less than a modern degree in some specialized medical area. No assertion or claim of a gift went untested. A common method of accreditation was a public feast replete with songs, symbolic regalia, and appropriate dances. Anyone who claimed to be what they were not was soon brought to public shame and consequently lacked public confidence. There were, and always have been, charlatans as there are in every society. But for the most part, the vast store of rattles, masks, head dresses, whistles, boxes, totem poles, house posts, and body designs recorded by anthropologists are testament to the authenticity of lived experience of indigenous people with spiritual power. Each item symbolized a tried and tested way of accessing power in the spiritual realm and this was common practice.

Charles Darwin, making a few observations over a relatively short period of time, developed a theory of evolution that created a new type of origin story. In this story everything in the beginning was simple and, over a long period of time, things gradually became complex until everything culminated in the most advanced species known as Western man. Indigenous people were not included in the end product of Darwin's evolutionary process but were assumed to be primitive examples of evolutionary genetic ascendancy. Biodiversity just happened by chance in this process. Although Darwin's idea is no longer the only theory of evolution, its continued prevalence has serious implications for the relationship between all life forms.

The first serious implication is the absence of values (such as respect) in the evolution story. The development of life forms is neither sacred nor profane; it just is. Reality is neither more nor less than empirical. According to the Tloo-qua-nah principle, however, when there is an absence of deliberate struggle for balance and harmony, the natural outcome leads inevitably to imbalance and disharmony. The natural outcome is an absence of sustainability.

The different perspectives presented in this paper are intended to draw broad strokes upon a complex subject. Neither the Western nor the indigenous perspectives are comprehensively presented. Nevertheless, the broad strokes point to fundamental differences that reflect actual Reality. While spirituality is not unknown in the Western world, it is marginalized from the centers to the periphery of power and decision-making. Science and technology are central to the Western world and the governance systems uphold democratic principles that focus upon human relationships. The *Canadian Charter of Rights and Freedoms* and the

American Constitution both assume the primacy of relationships between humans with no reference to other life forms. On the other hand, in indigenous societies, such as the Gitksan, the Wet'suwet'en, and the Nuu-chah-nulth, the relationship between all life forms is the very foundation of laws.

It is now these ancient indigenous laws that relate humans to every other life form that make up the contemporary issue of sustainability. The power to create a balance and harmony between all life forms rests with the Western world. Traditional indigenous people have spoken for the salmon, bear, eagle, cedar tree, and wolf, that these should be recognized and respected, but can the Western world respond effectively soon enough?

References

Brown, J.E. (1986) *The Spiritual Legacy of the American Indian.* Crossroads Publishing Company, New York.

Bunnell, F. and Atleo, R. (1995) *The Scientific Panel for Sustainable Forest Practices in Clayoquot Sound: Report 3: First Nations' Perspectives Relating to Forest Practices Standards in Clayoquot Sound.* Cortex Consultants Inc., Victoria.

Gisday Wa and Delgam Uukw (1992) *The Spirit in the Land. Statements of the Gitksan and Wet'suwet'en Hereditary Chiefs in the Supreme Court of British Columbia, 1987-1990.* Reflections, Gabriola.

Knudtson, P. and Suzuki, D. (1992) *Wisdom of the Elders.* Stoddart Publishing Co. Limited, Toronto.

Sam, S. (1992) *Indian Doctor Stories.* Transcribed by the Scientific Panel for Sustainable Forest Practices in Clayoquot Sound. Unpublished manuscript.

PART IV

Theories Relating Aesthetics
and Forest Ecology

Chapter nine:

An Ecologist's Ideas About Landscape Beauty: Beauty in Art and Scenery as Influenced by Science and Ideology

Daniel B. Botkin
Department of Ecology, Evolution, and Marine Biology, University of California, Santa Barbara

1 Introduction

The beauty of natural landscapes is one of the main justifications put forward for the conservation of nature, including the conservation of old-growth forests and other pristine natural areas. People find new clearcuts ugly, and one major opposition to clearcutting as a forest practice stems from a desire for landscape beauty. Stated most simply and in an extreme form, a common modern premise about landscape beauty in forested environments is that pristine forests, untouched by human beings, are beautiful, while in contrast the more intense the human action on forests, the uglier the landscape. This premise is reflected, for example, in the predominant visual resource management (VRM) strategies of the US Forest Service over the last quarter of a century (USDAFS, 1974), as discussed by Sheppard in this volume. However, the issue is more complex than this simple statement. For those not familiar with the development of ecology and other environmental sciences in the 19th and 20th centuries, some background on the possible underpinnings of this premise might be helpful.

The idea of pristine nature as a constant, and in a steady-state, dominated ecological thought until the last quarter of the 20th century. The prevailing ecological theory of that time, expressed in mathematical equations about the growth of populations and in many descriptions of the development of vegetation communities, assumed that populations, communities, and ecosystems established a steady state or climax condition (Botkin, 1990). The steady-state condition for ecological communities and ecosystems was believed by scientists to have the greatest biomass, greatest biological diversity, greatest stability, and greatest persistence. It was believed to be the best condition for both nature and people.

There were many practical consequences of this prevailing interpretation of nature. Laws and international agreements developed in the last quarter of the 20th century generally accepted the steady-state premise. This assumption is believed whether the goal is the harvesting of a specific species, such as a fish or a tree species, or whether the goal is the conservation of nature. Thus, in fisheries, formal mathematical equations calculated a *maximum sustained yield* based on a steady-state population growth equation. Forests were harvested according to a site index that was taken to be a permanent attribute of a habitat.

Accompanying this ecological perception of nature and the laws and international agreements, there was the development of what one might call a *natural aesthetic*, which assumed that what is natural is beautiful, and what is natural is nature's steady-state, *without human influence*. This aesthetic is in contrast to some important historical precedents of landscape beauty that existed outside of the concerns of ecologists and, for one reason or another, did not seem to override the continuing popularity of this *natural aesthetic*.

Today, as a result of major discoveries in environmental sciences in the last quarter of the 20th century, it is widely recognized by scientists that ecological systems are non-steady-state; that they are characterized by change, rather than consistency. However, the consensus is not universal, and one still finds in ecological theory, in ecological papers, and in the application of ecological ideas to the conservation of nature, a frequent reliance on the steady-state idea, sometimes implicitly, sometimes explicitly. In the United States, steady-state ideas still dominate laws and policies about the environment (Botkin, 1990).

The pressures against timber harvesting methods such as clearcutting, regardless of the specific conditions, can be seen as part of an ideology that began with the development of ideas about steady-state nature within ecological science, and, although shown to be inaccurate, these ideas still have a hold on the general public and many conservation organizations.

It is therefore useful to consider what a modern ecologist - one who accepts the idea of ecological systems as non-steady-state - might see in the recent history of landscape beauty appreciation, as it relates to prevailing ideologies and on-the-ground experiences of practicing ecologists and foresters. It is instructive to explore the representation of nature by landscape painters, and the reputation and admiration of specific works of art, as a kind of evidence about what people actually find beautiful in landscapes. It can also be instructive to consider anecdotal impressions by the public about the beauty of forested landscapes, and experiences of modern naturalists working in forests.

In this chapter, I attempt a very personal assessment, recognizing that art history and the history of landscape architecture are not my expertise, and that there has been a long history of concern by artists, designers, historians, geographers, psychologists, and philosophers with these issues (*e.g.* Carlson, 1977; Meinig, 1979; Claik, 1986; Appleton, 1996; Kaplan *et al.*, 1998). My purpose is not to discount their work or simply replace it, but to try to add to the discussion a new perspective that arises from what has been called the *new ecology*. What I have done is to look, through an ecologist's eyes, at how painters have portrayed nature, making use of my experience in, and knowledge of ecology. In doing this, I am treading on dangerous ground, because I am not a professional art historian or critic nor have I any ability as an artist myself. However, I have long been interested in the representation of nature in art, both because I love graphic arts and because I find paintings to be representative of their times in perceptions of the beautiful. I believe that the history of the artistic rendering of nature is useful when we ask what is, in fact, aesthetically pleasing. In this chapter I discuss specific works of art and give an interpretation about what these works of art can tell us about what people actually find beautiful, aside from an ideological position inferring what they *should* find beautiful.

This is not to suggest that landscape painters have been or are unacquainted with the realities of nature; quite the contrary. In general, those who examine nature in detail achieve an understanding and empathy with her that is greater than those who do not. One might therefore expect landscape painters in general to be more in touch with and understanding of nature than their public. This chapter focuses on an ecologist's perspectives on landscape beauty; other chapters in this volume focus on the many other symbolic and ideological meanings that are important in landscape appreciation in the modern world (for example, see Harrison, 1992).

2 Appreciation of Human Influence on Nature

Consider the question whether heavily disturbed, cleared forests could ever be considered beautiful. Imagine a landscape that had been forested, but had been logged a long time ago. Afterward, people removed every sapling that appeared there, so that the landscape was maintained deforested. If one were to take what I have called a *natural aesthetic* to its logical conclusion, one would have to believe that such a scene should not, by definition, be beautiful. The scene I have just described is portrayed in Van Gogh's famous painting, *The Harvest* (Los Angeles County Museum) (Figure 9.1; Plate 4). This scene is of a landscape that would revert to a temperate forest, if it were not maintained in agriculture. The popular appeal of this painting was brought home to me recently at the Los Angeles County Museum, where I saw a traveling exhibit from the Van Gogh Collection in Amsterdam. At $17.50 a person (U. S.), there were 600 to 700 people per hour passing through the exhibit, all trying to view Van Gogh's paintings including *The Harvest* and others of French agricultural landscapes. Based on the judgement of art historians and art critics, or based on the money people are willing to spend to view these works, it is hard to argue that the landscape portrayed in *The Harvest* is ugly. This presents an apparent paradox. According to common modern ideologies about forested environments, people should be revolted by this treatment of the landscape, but they are not. This is in obvious conflict with the premise that *nature is only beautiful when free of human action, and ugliness increases with human action*, and with a less extreme view that *ugly landscapes tend to be man-made, while beautiful landscapes tend to be free of human influence*. Of course, it is quite possible that many of the viewers of Van Gogh's *The Harvest* are not aware of

Fig. 9.1. Vincent Van Gogh. *The Harvest* (see plate 4). Source: Los Angeles County Museum.

the ecosystem degradation that followed the deforestation in the scene, but even if they were, there is ample evidence of people's attraction to productive farmland (see, for example, Nassauer, 1995, and Sheppard in this volume).

The apparent paradox deepens when we consider real world examples, such as the popularity of the view from the summit of Mount Monadnock in New Hampshire. Monadnock is an easy drive from Boston and a popular destination; a hike to the summit takes about two hours. The summit provides a beautiful view of the New England countryside, but the mountain is being loved to death. Overuse as a recreational site has become an environmental issue; today, the summit is exposed bedrock completely barren of trees and shrubs; vegetation is represented by lichens and mosses (Rosenthal, 1999).

In the early 19th century, Monadnock was forested to the top, but dense forests in the valleys prevented New England settlers from farming the land. There were more trees than the people could use, and, needing to grow food more than they needed to look at forests, settlers cleared the land using the most efficient methods available at the time. First they killed the trees by girdling them (cutting through the bark around the main trunk). Then the settlers burned the forests. Not caring about the mountains, they did nothing to stop the fires, which burned over the top of Mount Monadnock. The fires were so hot that they burned away the organic mat that formed the soil on the summit. It is said that these fires were hot enough to crack the granite bedrock of the mountain. I believe that few people who hike Mount Monadnock have any idea that it was ever forested to the top. If they did would they find the view less beautiful?

Van Gogh's *The Harvest* and the summit of Mount Monadnock illustrate deforested landscapes that are considered beautiful. Another Van Gogh painting, *Courting*, illustrates another aspect of the paradox on perceptions of human influence in the landscape: can young plantations - artificially planted and managed stands of young trees - be considered beautiful? Much opposition to forest plantations has been expressed in parts of the world such as the United Kingdom and Australia. However, *Courting* suggests a different view. It is a painting of young trees planted in straight rows and maintained within a park, but it is a painting of a great beauty in the ordinary sense of that word. It suggests that plantations can appear beautiful, depending on their design and context.

That a forest plantation in reality can be taken as beautiful was illustrated in a recent visit I made to Plum Creek forestland in Maine. Carl Haag, a forester for that company, took a group of ecologists and foresters on a tour of their land practices. One of the places we visited was a mature plantation of spruce and white pine. That it was a plantation was clear from the regular spacing and uniformity of the sizes of the trees. The spruce had grown large enough to form a canopy that shaded the ground. The ground was covered with brown needles and a sparse undergrowth of some shrubs. Haag said that he brought a group of tourists to this location and that one of them refused to believe that this was a plantation, *because* it was beautiful. Here, ideology about the effects of human beings on nature prevented this tourist from believing that people could ever take actions within a forest that produced beauty. Finding the plantation beautiful, the tourist therefore was convinced that it must be free of human influence. This is not to imply that all tree plantations are beautiful, but it does suggest that landscape design can make all the difference in the aesthetic quality of a plantation or any managed forest.

3   Historical Shifts in Landscape Appreciation

The idea of the beautiful in landscapes has changed during the history of civilization. In recent centuries, scientific discoveries have had a significant effect on the idea of the beautiful in nature. One major change in the idea of landscape beauty is illustrated, for example, by the landscaping at the Rockefeller-Bellagio Institute on Lake Como, north of Milan. The Institute is housed in a large mansion whose earliest structures date to 1492 and which were added to over the centuries.

The landscape in the grounds of the Institute has two sections. Below the main building are classic Italian gardens; up the hill from it is an English garden. The Italian garden at Bellagio has shrubs and trees that are trimmed into regular geometric shapes, and are planted in geometric formations. That Italian garden demonstrates one common aspect in Greek and Roman gardens two thousand and more years ago: the idea that symmetry is an important component of landscape beauty. It was common in the classic world to believe that, if nature was not symmetric, then the proper approach was to revise it and make it symmetric.

The English garden, a much later development which includes what has been termed the picturesque (Porteous, 1996) and that we today might call naturalistic or natural-like, represents an entirely different idea of landscape beauty. A walk through the English garden is a walk through young shrublands and woodlands, made to appear quite natural in the sense that individual plants are not trimmed to geometric forms, but appear rambling; they are not planted in geometric formations, but grow as in a young forests, with many overlapping stems of different species, with thickets and clearings. The vegetation obscures much of the view most of the time, so that as one walks through the garden one is not sure what lies ahead. There is a sense of mystery and discovery. At a turn in a path, a visitor suddenly comes upon a statue and a park bench, at another turn, a gap in the vegetation reveals a view of the Italian Dolomitic Alps rising about the blue waters of Lake Como.

In the classic world, such a view of the Alps would have been considered ugly. The Greeks and Romans wrote that mountains were the warts on the Earth's surface: without symmetry, they were not beautiful. While there were other classical "rules" for beauty, such as the *golden mean*, ideas of beauty in nature were sometimes so focused on symmetry that philosophers wrote that there had to be a place in the ocean whose depth exactly equalled the height of the tallest mountains. In that way, the apparent asymmetry caused by the existence of mountains would be balanced and the world, as a whole, would maintain its symmetry (Botkin, 1990).

The evolution from the classic Italian gardens to the English gardens represents a transition in the idea of beauty that took place during the scientific and industrial revolution, beginning with changes in knowledge about the Cosmos and in the understanding of the physics of mechanics – the result of work by Newton, Kepler, and Galileo and Copernicus (*Encyclopedia Britannica*, 2000).

Galileo was the first person to use the telescope in a methodical study of the Cosmos. In the early 17th century he collected evidence about the movement of the stars and planets that showed that the Earth revolved around the sun; that the surface of the moon was rough and irregular rather than smooth; and that the sun's surface had dark spots. The telescope revealed a Cosmos that was not structurally symmetric and therefore not perfect in the classic sense.

Kepler contributed to the change in the conception of the Cosmos, demonstrating that the planets moved in elliptical rather than circular orbits. Because a circular orbit had been believed to be the perfect shape, Kepler's findings seemed another blow against the structural perfection of the Cosmos. Furthermore, Kepler's and Galileo's work proving that and that the Earth was not the centre of the universe represented a blow to the belief that the universe consisted of God-made-perfect, human-centred structures. These findings seemed to threaten one of the classic arguments for the existence of God: God, being perfect, had to create a perfect world, and perfection was believed to exist in physical structure. The new ideas seemed heretical.

The resolution between science and Christianity came with the development of Newton's laws of motion. These suggested that there was a new kind of perfection in the Cosmos, a perfection of processes, of motion, and dynamics. God was revealed by Newton's laws of motion to have created a Cosmos that was perfect not in structure, but in these underlying laws that governed motion. This is a kind of perfection at a conceptual level more subtle than that of structural perfection. Accompanying this change in the understanding of the Cosmos was a change in the perception of landscape beauty. As explained in a classic work on the subject, *Mountain Gloom and Mountain Glory*, by Margorie Nicolson (1959), the transition was from a conviction that beauty lay only in static physical structure to an appreciation of the power of nature - taken as a representation of the power of God - as illustrated by storms at sea, rugged mountains, and wilder, naturalistic English gardens.

From a religious perspective, there was a transition from the view that God had made a world that was statically perfect, and therefore must be symmetrical, to a belief that God had made a world in which Newton's laws of motion ruled, and they had their own beauty. So, process became beautiful. A crucial period in this transition occurred during the late 18th century and early 19th century. The beginning of the transition is represented by Edmond Burke's *A Philosophical Enquiry into the Origin of our Ideas of the Sublime and the Beautiful*, published in 1775. He distinguished between the beautiful and the sublime, terminology that became common among the Romantic Poets.

Prior to Burke, in the early Renaissance, a trip through the Alps was considered horrible. By the end of the 18th century, a traveller wrote that climbing the Alps, he experienced a *horrible joy*, which suggests the beginning of an appreciation of mountain beauty. Soon after, a hike through the Alps became a popular recreation for those who could afford it. The Romantic poets further developed the distinction between two kinds of pleasing landscapes: the beautiful and the sublime. For the classic symmetrical landscape, they reserved the word *beautiful*; for awe-inspiring experiences of the power of nature, represented by the Alps and by the ocean, they reserved the term *sublime*. This transition occurred in painting as well, as illustrated by Frederick Turner's painting of a steamboat in a storm at sea, which shows a dynamic, powerful and dangerous ocean portrayed as an object of beauty. There was a connection between scientific discoveries, religious beliefs and ideas of landscape beauty (Botkin, 1990).

At the beginning of the 19th century, Meriweather Lewis was aware of the transition taking place in the idea of beauty in landscapes. When Lewis arrived at the great waterfalls on the upper Missouri - at the location of modern Great Falls, Montana, he struggled to write a description of this beautiful landscape (Botkin, 1999). At that time, the Lewis and Clark expedition was at a crucial point. Lewis and Clark had to get their men and equipment up beyond the falls, up the rest of the Missouri River, locate Indians from whom they could buy horses, and find their way over the Rocky Mountains before winter. Lewis understood the precariousness of their situation, but he was so struck by the beauty of the falls that he stopped what else he was doing and spent two days doing nothing but looking at these falls and describing them in his journal. Previously, most of Lewis' journal was a direct and objective reporting of what he saw. Lewis wrote about major events and discoveries, such as the description of a new species. At Great Falls, however, Lewis revealed his emotions, his artistic sensitivity, and the influence of his education.

At this first set of falls, Lewis saw a rainbow in the spray as the sun reflected off the water. This, he wrote, "adds not a little to the beauty of this majestical [sic] grand scenery". He sought within himself an ability to express the beauty of the landscape. "After writing this imperfect description I again view the falls and am so much disgusted with the imperfect idea which is conveyed in the scene that I determine to draw my pen across it and begin again, but then reflected that I could not perhaps succeed better," he wrote. He wished for "the pencil of Salvator Rosa" a seventeenth century Italian landscape painter of wild and desolate scenes, and for "the pen of Thompson," an 18th century Scottish poet who was one of the forerunners of the Romantic Movement. On June 14, 1805, Lewis reached several more of the falls and was most impressed with one he called Rainbow Falls, which is now much altered by Rainbow Dam. This is "one of the most beautiful objects in nature," he wrote. Lewis spent some time trying to decide which of the two, the falls he had seen the day before or this one, was the most beautiful. "At length I determined between these two," he wrote, that Rainbow Falls was "pleasingly beautiful" while the one he saw the day before was "sublimely grand" (Botkin, 1999).

These are the turns of phrases I mentioned earlier that were in use among the Romantic poets to describe aspects of beauty. Lewis was using phraseology that would have been familiar in the aristocratic drawing rooms of England, and in Jefferson's Monticello mansion. This distinction would have been unlikely to have occurred to other explorers of the American West in Lewis' time or for some decades after. Contemporary and evolving ideas of landscape beauty affected Lewis when he found himself within a vast wilderness, unseen, to his knowledge, by any person of European descent. His experiences illustrate the power of ideas about landscape beauty and the power of landscapes to affect a thoughtful and observant person.

4   Appreciation of the Dynamics of Nature

In the early nineteenth century, a school of landscape painting developed that is known in North America as the Hudson River School. A classic example of this style is Church's painting called *Morning in the tropics* (National Gallery of Art) (Figure 9.2; Plate 5). As an ecologist, it seems to me that the Hudson River School

Fig. 9.2. Church. *Morning in the tropics* (see Plate 5). Source: National Gallery of Art.

paintings typically had three characteristics that are important to our discussion of the connection between scientific understanding of nature and the idea of the beautiful in nature. Paintings such as *Morning in the tropics* depict nature as very beautiful; completely static; and having either no people in it, or people dwarfed by nature. These three classic characteristics persist more generally today in the appreciation of forests and in resource management, and underlie some common modern attitudes about what is good and what is beautiful in nature. Some of the work of the Hudson River School therefore can be seen as an early representation of a dominant modern view of landscape beauty, to which I have already referred: nature is good and beautiful without people and without human influence, or perhaps one would say nature is best when human influence is least. Ideas about landscape beauty along with the portrayal of the beautiful in nature have had political consequences. Later in the 19th century Thomas Moran painted some of the great scenery of the American West, including his famous painting of the Grand Canyon of the Yellowstone. Moran popularized the awe-inspiring scenery of the American West to the point of probably helping the movement that created American national parks.

One of my favourite landscapes is from a hilltop in Alstead, New Hampshire, where one looks west to the Connecticut River valley. On a clear autumn day, one sees all of the lovely colours of autumn: the yellow leaves of white and yellow birch, early successional species that come in after a clearing and do poorly in the deep shade of an old forest; the green of white pine that only germinates and survives after a clearing; the orangish hue of red oak and the fiery red of sugar maple - species characteristic of old-growth New England forests. The beauty in this mixture of colours is the result of a landscape mosaic formed of different stages in ecological succession present on the land at different locations at the same time. If the landscape were only old growth or only early forest, it would be less beautiful. Such scenes have been the subject of many a painting and photograph. It is a visualization of a dynamic nature because it requires all the stages in ecological succession, in a process, not in a static scene. Over time, the location of the different stages will change; but as long as there is a sufficient rate of clearing at sufficient intervals, the landscape will be sustainable and beautiful.

This situation suggests that there may be a resolution to the supposed paradox I presented at the outset of this chapter. The paradox arises from the premise that, according to prevalent modern ideas, forested nature is most beautiful when left free of human action, and the more intense the human action, the uglier the landscape; while the examples from art and from forestry practice suggest that people find many forested and previously forested landscapes that have been greatly altered by human actions to be beautiful. More often than not, they find these beautiful when they remain unaware that the landscape has been altered by people or perhaps that the landscape has been altered in culturally appropriate ways (see Chapters 11 and 12, this volume). Sometimes, as in the case of the visitor to the Plum Creek Plantation, the blinders of ideology led a person to believe what is clearly not true. The resolution may lie in accepting the essential dynamic character of biological phenomena and processes. Once change is accepted as "natural" it can then perhaps be accepted as beautiful, at least in certain circumstances.

5   The Power of Ideology in the Appreciation of Nature

The previous ideas on resolving apparent paradoxes in landscape appreciation and the understanding of nature in forested environments suggests that the more we can free ourselves from ideology, the more directly our senses may respond to the beauty of a wide variety of landscapes. However, under the influence of powerful ideas, of powerful ideologies, what is acceptable as beautiful becomes restricted. This is not a new idea – over twenty-five years ago, Meinig (1979) explored thirteen different viewpoints on the same landscape scene, as viewed by different professionals. It is therefore clear that what is accepted as beautiful is not independent of the intellectual perspective of a time in the history of civilization, strongly influenced by the power of ideas. Simultaneously, what is accepted as beautiful affects perceptions of nature that in turn influence ideas about the character and workings of nature. A relevant example of the problem in practice, even among reputable scientists, is a study done some years ago by an ecologist from Massachusetts, John M. Hagan III, Director of Conservation Forestry at Manomet Observatory, Manomet, Massachusetts (Botkin, 2000). Hagan became interested in the Maine Woods. In *The Maine Policy Review*, he wrote a kind of *mea culpa* essay, for which one can only praise him for his honesty and willingness to share what he learned. Hagan wrote that he saw the clearcutting issue as a "bitter debate over jobs for Maine people versus the future of the North Maine Woods." He told his own story, beginning with his initial interest in conducting a research project about the possible effects of Maine forest practices on the decline of migratory birds. Obtaining access and an introduction to the Great Northern Paper Company's land, he was taken to a clearcut (since this project began, the land has changed hands several times and is at the time of this writing owned by Plum Creek Timber Corporation). He was so appalled that he was "rendered incapable of simple conversation" and decided that his research plan would have "no need to survey the clearcuts" on the basis that "nothing could live in what I had seen, at least for many years." Here he had formed his conclusion *a priori*, without systematic observation, and without quantitative evidence. He had let his personal reaction to scenery (which I would suggest was influenced by and a consequence of modern ideology about nature), determine what he intended to claim was a scientific conclusion. Here Hagan violated the scientific method in the name of doing science, by forming an opinion which he had no intention to test, meanwhile intending to make what he believed was a scientific study.

Hagan's host on the Great Northern Paper Company's land was research forester Carl Haag. Haag and Hagan talked about the plan for research, and Haag persuaded Hagan that he should make measurements in clearcuts as part of doing a scientific study correctly. Hagan decided to follow Haag's advice on the basis that "once I had documented that clearcuts were avian deserts, public pressure might bring about kinder, gentler, forest practices."

Much to his surprise, however, when Hagan began his field work, he found the clearcuts "full of birds" including many that had concerned conservationists as threatened or endangered, such as the Chestnut-sided warbler and the American kestrel. To his credit, at this point Hagan abandoned his ideological prejudices and accepted the observations that confronted him. These observations led him to conclude that a set of forest stands of different ages, including clearcuts, was needed (Hagan, 1996). He recognized that a forest represents a process, not a fixed structure. Scientific observations changed his opinion. Once recognizing the need for the existence of a variety of stages in forest development, Hagan realized also that the issue had to be rephrased as: *what is the amount of disturbance or clearcutting that was necessary to sustain ecological processes?* Here he admits that his mind had been made up before he began his research, expecting that research to serve a political and ideological purpose. I admire Hagan for writing an article that sets forth his mistake and puts the record right. It is courageous, honest, and helpful. It is a breath of fresh air to read his article in this time of intensifying misunderstandings. We need to recognize that we all, even scientists, carry our ideologies with us in our perceptions of the environment.

6   Conclusions

From this brief examination of a few landscape paintings and real world examples of designed and managed landscapes, it is apparent that the idea of what is beautiful has changed over time, and that some heavily altered landscapes are taken to be beautiful. In the late 18th - early 19th century, Western civilization passed through a first, science-influenced transition in the idea of landscape beauty. I believe that a similar, second stage in this transition is taking place as a consequence of new ideas arising about biological nature from sciences that include ecology, geology, oceanography, and climatology. We are in the midst of this transition, and do not know yet what is the appropriate terminology. Ecology remains a young science that as yet has few solid answers to complex environmental problems. Assuming that this science continues to advance, one would also expect additional implications for both scientific and lay-public perceptions of forests and landscape beauty. In the meantime, among forest managers and ecologists charged with stewardship of the land, there is likely to be confusion about what are the most appropriate rules for what is beautiful in biological nature.

Another conclusion of this review of art and scenery is that in naturally forested areas, old-growth is not the only kind of scenery considered beautiful. The famous New England autumn landscape presents a mosaic of many stages in ecological succession, from farmed openings to mature, but not pristine, forests. In such situations, therefore, an aesthetic argument is not by itself sufficient to justify the prevention of harvests, if maximizing scenic values is a priority.

I want to be completely clear that a conclusion that *does not* follow from my discussion is that we can treat landscapes any way we want, and that no matter what we do people will find it beautiful. This chapter is not meant to be an apology for excessive or poorly designed clearcuts, but instead an examination of what people have found beautiful in nature and why. An implication of this chapter, consistent with Nassauer's findings (1995) is that careful use and well-cared for land, such as the farmland depicted by Van Gogh, can be appreciated as beautiful. To this I would add that some heavily altered landscapes, ones cared for like a well-run farm, or like a careful designed park, can be seen as beautiful. These landscapes can be consistent with the modern idea of the sustainability of nature and natural resources. For example, the French farmland has been maintained as such for many centuries. A park created from plantations can be sustained. Some of these ideas are discussed in more depth by Sheppard in this volume.

It is essential also to make clear that it is *not* my purpose to use *The Harvest* and the other paintings and scenery to justify destructive, non-sustainable use of the land. While people can come to appreciate that processes and dynamics can be beautiful, one also needs to recognize the symbolic and aesthetic properties that ancient forests such as stands of giant sequoias and redwoods have for many people, and the sense of loss that people feel from the cutting of these ancient forests. The scarcer such scenery becomes, the greater the sense of loss and the greater value people will place on the beauty of scenery.

Finally, the scientific transition that began with physics and astronomy needs to be completed in the next, biological century. This places a burden on ecological scientists. We need to communicate the idea about the dynamics of biological nature, to the public and to artists (as well as poets and novelists). These may in turn lead to new ideas of what is beautiful in landscapes, and to a mutually improved understanding of landscape values.

References

Appleton, J. (1996) *The Experience of Landscape*. John Wiley and Sons, Chichester.

Botkin, D.B. (1990) *Discordant Harmonies: A New Ecology for the Twenty-First Century*. Oxford University Press, New York.

Botkin, D.B. (1999) *Passage of Discovery: The American Rivers Guide to the Missouri River of Lewis and Clark*. Perigee Division of Penguin-Putnam, New York.

Botkin, D.B. (2000) *Nobody's Garden: Thoreau and a New Vision for Civilization and Nature*. Island Press, Washington, DC.

Burke, E. (1988) *A Philosophical Enquiry into the Origin of Our Ideas of the Sublime and Beautiful*. A. Phillips (ed), Oxford University Press, New York.

Carlson, A. (1977) On the possibility of quantifying scenic beauty. *Landscape Planning* 4: 131-72.

Claik, K.H. (1986) Psychological reflections on landscape. In: E.C. Penning-Rowsell and D. Lowenthal (eds) *Landscape Meanings and Values*. Allen and Unwin, London.

Encyclopedia Britannica (2000). <http://www.Britanica.com>.

Hagan, J.M. (1996) Clearcutting in Maine: Would somebody please ask the right question? *The Maine Policy Review* July, pp. 7 -19.

Harrison, R.P. (1992) *Forests: The Shadow of Civilization*. University of Chicago Press, Chicago.

Kaplan, R., S. Kaplan and R.L. Ryan (1998) *With People in Mind*. Island Press, Washington, DC.

Meinig, D.W. (ed) (1979) *The Interpretation of Ordinary Landscapes: Geographical Essays*. Oxford University Press, New York.

Nassauer, J.I. (1995) Messy ecosystems, orderly frames. *Landscape Journal*, 14(2): 161-171.

Nicolson, M. (1959) *Mountain Gloom and Mountain Glory: Development of the Aesthetics of the Infinite*. Cornell University Press, Ithaca, New York.

Porteus, J.D. (1996) *Environmental Aesthetics*. Routledge, London.

Rosenthal, D. (1999) Environmental issue. In: D.B. and E.A. Keller (eds) *The Earth as a Living Planet* (3rd edition), John Wiley, New York.

United States Department of Agriculture Forest Service (1974) *National Forest Landscape Management, Volume 2*. USDA Agriculture Handbook No. 462. US Government Printing Office, Washington, DC.

Chapter ten:

Can a Fresh Look at the Psychology of Perception and the Philosophy of Aesthetics Contribute to the Better Management of Forest Landscapes?

Simon Bell
School of Landscape Architecture, Edinburgh College of Art/Heriot-Watt University
Edinburgh

1   Introduction

This chapter looks beneath the types of perceptions and preferences expressed by the public about landscapes in general and about forest landscapes and forest management in particular. Past research tells us much about what kinds of landscape people like, but does not necessarily explain why they like it; much research on the forest landscape has concentrated on the landscape as scenery viewed externally. Attempts to manage landscapes with this kind of aesthetics in mind often conflicts with other factors, unless integrated well.

   This chapter returns to some of the theory underlying perception and aesthetics, and attempts to assimilate the two fields of study in order to discern some useful principles that can be applied in research and practice. The chapter is arranged in two main parts. The first section commences with a brief look at the mechanisms of perception, and some of the psychological processes at work, concentrating on how these are used by us to make sense of the world. The second section applies these conclusions about perception and applies them to a number of facets of aesthetic philosophy, reviewing some concepts that reveal possible fresh insights into the importance and applicability of aesthetics. Finally, the implications of this for the management of forest landscapes are suggested.

2   Perception: The Starting Point for the Aesthetic Experience

This section summarizes the main components of perception in order to provide a basis for discussions to be developed later in the paper. For a fuller explanation of the subject see Bruce *et al*. (1994) and Bell (1999). This section will consider the way the senses are used in perception, the mechanisms of visual perception, some psychological aspects in terms of how we understand the world and the process of pattern recognition as a starting point for the aesthetic experience. In this chapter, I refer to perception as the process for gathering and processing information about landscapes, as distinct from aesthetic responses and preferences which can arise from the perception process.

2.1 The Senses and Their Role in Perception

It has long been traditional amongst aestheticians and philosophers to divide the senses into those dealing with distance and those for nearness, or the proximal. Sight and hearing are usually thought of as the distance senses. These have also been identified since the time of Kant (1724-1804)as the aesthetic senses, because they allow us to reflect on a scene, art object or music from a distance. The other senses of touch, smell and taste have been relegated to more utilitarian roles.

Sight uses light energy to detect shapes, textures, colours, and intensities of light and movement, together with aspects of spatial dimensions such as distance and depth. It is a particularly important sense for humans, one to which we have become evolutionarily highly adapted, forming one of the main ways in which we think and picture concepts in our head.

Hearing uses the energy of sound waves, as noise, or as specific patterns of pitch (high to low notes), order, rhythm, beat, timbre (the quality or character of the sound). We may recognize some patterns of sound as unique, such as the sound of running water or the voice of a person well known to us. We can use hearing to reinforce the information provided by sight.

Smell uses the nose to detect small amounts of chemicals borne on the air and is one of the more primitive senses in terms of its evolutionary development. It is also immediate in its effect and very powerful at triggering the retrieval of complete memory sequences, from apparently simple odour combinations. Taste depends on putting substances into our mouths. We can taste four sensations: sweet, sour, salt and bitter. When taste and smell are combined, flavour is obtained.

Touch or tactile sensations are much wider in their scope than is often imagined. With the basic sense of touch we use mechanical energy to feel shape, texture, and pressure. Using other sensory cells in our skin, we can also feel temperature, humidity and pain. Our body movements tell us about direction, elevation and the degree of resistance offered by surfaces underfoot. These tactile senses can be grouped together under the general heading of haptic or *kinaesthetic*, meaning perceived through movement. They are all proximal in their application; that is we detect stimuli from them directly.

All these senses are usually interconnected and this is important in giving us a complete picture of our environment. However, some are stronger or more important than others under different circumstances. When considering "landscape" in the sense of a "prospect of scenery", the visual impression is by far the most important for perception and the corresponding aesthetic response. There are also important distinctions to be made between the landscape as scenery and the environment as a multi-sensory engagement, where the aesthetic response is different; this is of particular importance when we are in a forest as opposed to looking at it from a distance. These differences are significant when considering certain key aesthetic approaches as discussed later in the chapter.

The amount of information and knowledge we obtain depends on the variety and degree of contrast of the sensory data that we receive and the extent to which we can differentiate them: that is, to detect patterns. This aspect of pattern recognition is a major link between perception and aesthetic response.

Obtaining and using sensory input about the world is much more than mechanical reception of data, later processed by the brain as a separate activity. For example, at the same time as we perceive the world, we also project our subjective feelings and preconceptions onto it. This is why concepts such as "landscape" or "wilderness" are as much states of mind as they are physical entities; this has major implications for aesthetics and for the meaning of the term *environment*.

2.2 Mechanisms of Visual Perception

Visual perception starts with the reception of light into the eye. Light is emitted in different degrees from a range of objects and surfaces; it varies in wavelength (colour) and amplitude (brightness) and is affected by the medium through which it travels. The precise quantities and qualities of the light received by an observer vary, depending on the position. The totality of all the light received from all directions is known as the *optic array*. Changes in the optic array generally denote movement, either of the observer or of parts of the scene under observation such as animals, vehicles, clouds or water, although this is not always the case, such as when the overall light levels diminish. The detection of movement is an important aspect of perception.

The eyes sample the optic array by means of jumps, or *saccades*, between *fixation points*. The fixation is the short time (a few milliseconds) needed to gain an image of part of the array. Once an area of the array is fixed, pursuit, either of a moving object or of a static object by a moving observer, can be achieved to keep it in view. The constant saccades and fixes by the eye sample the optic array at high speed in order to build up an overall image of high acuity.

The eyes receive light of varying intensity and wavelength as an image on the retina, which is transmitted into the processing areas of the brain as a spatially related pattern. The brain has to interpret and make sense of that pattern and use the results to inform us how and where to look next.

Human eyes have evolved as aids to survival, so we undertake perception that is purposeful and selective. We need to understand the way the landscape is assembled, we seek out and notice change and movement, and we apply meaning where we wish to engage with our surroundings and feel part of the world we inhabit. We tend not to notice things that do not concern us. To that extent our view is subjective. Thus we have to disentangle the multitude of light patterns into something we regard as the reality of the scene and about which we can, with experience, predict certain cause and effect relationships. Various mechanisms and theories of visual perception have been proposed, as discussed below.

The Primal Sketch

The first things we are able to detect and classify in the landscape are the edges defined by varying light intensities (Bruce *et al.*, 1994). These begin the process of shape and pattern definition. One issue is to determine which edges belong to which shapes and thus to separate out the elements of the scene. In addition, changes in viewer position, relative to the light source, will yield a completely different set of edges from the same set of elements or objects.

In the absence of biological models or a clearer understanding of the cellular level processes, psychologists, in particular Marr, have devised their own models or algorithms to deduce the processes of perception that allow us to make sense of the scene (Marr, 1982). One of these is the concept of the *primal sketch*, the first transformation of the retinal image into a derivative that describes the detailed features of that image. This sketch has to be able to define the edges of the image from the light intensity and to compute their significance and general associations with different elements of the scene.

This sketch is synthesized from three derivatives: the steepness of gradients of light intensity across the image, variations in energy intensity in different parts of the image and the way these trigger responses by several types of brain cell at various thresholds (Marr, 1982). Concepts of *signal processing* are involved, similar to the way radio signals are received; the strongest signals occur where intensity variations are more extreme.

According to this theory, the retinal image has been reduced to its basic elements: a set of edges, contours and energy variations across the scene. It is important to know which are linked and how they are combined into more complex images. Early concepts used ideas of *association*, of linking like with like, or with simpler images combining to form more complex ones (Bruce *et al.*, 1994). These concepts do not explain how shape and order (the way elements are organized to form structures) are achieved when these elements overlay any existing associations.

Gestalt Psychology

One of the most influential set of theories to explain how more complex images are created from the identification of edges and contours in the primal sketch, is known generally as *gestalt* psychology. "Gestalt" means, in German, both *shape* and the *character of the shape*. Kohler and others developed ideas about how we determine shape, together with a range of other spatial cues, in order to discern order in the images presented to us (Kohler, 1947).

The first of these spatial cues is commonly referred to as *figure and ground* where the element being identified stands out as a separate entity or "figure" from the rest of the scene, or at least its immediate surroundings. The perception of a figure can depend on its shape, its colour, texture and position in relation to the background when viewed in three dimensions. In addition, there are several other "gestalt laws of organization", which are all spatially based. Proximity or *nearness* of visual elements causes them to be perceived as a discrete group, especially when they are similar in shape, colour, texture, direction or position. In addition, the *common fate* of such elements, in terms of how they move in a recognizable pattern, can also help to distinguish their character. Amongst these laws are *continuity* of patterns, such as the position of elements into a line which can be read at a sweep, or *closure*, where there is enough evidence to suggest a shape, from elements partly enclosing space.

Active Perception

There is evidence that different people will look at the same scene but perceive different shapes and patterns depending on their knowledge, experience, culture and so on; this further reinforces the theory that active, selective and intelligent perception is normal, as opposed to passive sampling (Bruce *et al.*, 1994). The greater the involvement by the observer in the landscape, the greater the degree of intelligent perception and active visual thinking that occurs.

In many instances, intelligent perception acts as a filter to determine what is worth seeing and comprehending in an otherwise confusing scene. Looking for a well-known face in a large crowd involves having the image as a template, so that key features are sought from all the possible faces present. As soon as the person is seen, the selective cognition takes place, thus saving the great effort of scanning each face separately. This phenomenon explains why different people see different things in the same scene: to some extent, they are preconditioned to look for patterns they can recognize. It may also contribute towards the connection between perception and aesthetics to be developed later in the chapter.

Gibson's Theory of Affordance and Optic Flow

In the discussion so far, the psychology of perception has been based on the processing and comprehension of the optic array, starting with the retinal image. We rarely see objects without a background that is three dimensional and composed of surfaces, which have their own qualities of texture, colour, density, form, orientation and so on. This perception of surfaces at varying distances is important, because it gives a framework for the space being viewed.

Depth perceived by the stereoscopic image of an isolated object also includes the qualities of the surface it sits upon, such as the gradual change to texture due to the proportionately decreasing size and interval of the elements that comprise the surface (such as gravel or tiles) with increasing depth. The actual scale of the visible landscape is relative to the known size of the observer and the intervening distance, whereas the perceived size of a person seems to be the same whether they are close or far from us, despite the decreasing size of the retinal image. Thus we perceive variable information, yet relate this to constant characteristics in a landscape such as the consistently receding ground surface or the position of the horizon relative to the observer.

This view differs from the theory based on the analysis of the retinal image and has been called the *ecological* approach (Gibson, 1966; 1979). This term is not the misrepresentation it first appears, because it treats visual perception as part of the necessary interaction between organisms and their environments. The landscape is viewed in relation to the viewpoint of a mobile observer, who is actively evaluating a moving optic array with respect to the information it might yield and not passively sampling a static optic array. This theory has some connection with the Gestaltists in one important aspect: that we view the world in terms of what various parts *afford* us in a utilitarian sense. For example, apples are to eat, a road is for driving along, a forest is for obtaining timber, for taking a stroll or for a wildlife habitat.

In many ways this approach only differs from that described earlier in terms of the added means of depth perception available to us. The more philosophical differences relate to the perception of movement and its vital importance to the whole system of perception. We constantly move our eyes, our heads, our bodies, our positions and elevations, and do this to build up a picture of the relationships amongst the surfaces and objects we are looking at.

As exemplified in the phenomenon of parallax, the automatically changing optic array follows some criteria that maintain the spatial relationships amongst the elements of a perceived scene. As we move into a landscape the scene flows towards and past us, whilst constantly maintaining a correlation between the changing optic array and our spatial position. This optic flow defines direction and is the mechanism by which we can judge distance and speed and so drive a car, fly a plane or shoot at a moving target. We make all these judgements as a continuous perception.

According to Gibson (1966), the perception of motion through our sensing of the changing optic array is considered to be a basic or primitive aspect of vision. In the same way that a number of the gestalt laws can be translated into terms used as design principles to describe a coherent pattern, so can the "Gibsonian" variables of texture, density and colour applied to surfaces.

The Gibsonian view of perception is most useful for animals and humans living and moving about the environment for various purposes. Navigation between and interception of other moving objects is a major part of this existence. In contrast, the design process has traditionally viewed the landscape as a static scene from fixed viewpoints, comprising the spatial relationship of several parts to the whole. This approach to perception, where the viewer is immersed or engaged in a scene, has a direct relationship to a particular branch of environmental aesthetics to be explored later in the chapter.

2.3 Processes of Pattern Recognition

Whether we subscribe to the psychological, Gestaltian or the Gibsonian views, the perception of patterns depends on two aspects: recognition of objects and their interrelationships. It is the relationships that are the key determinant for the pattern of a landscape.

Many of the qualities of pattern recognition can also be analysed in terms of the Gestalt properties, the development of the primal sketch or by optic flow. At this initial level of analysis, where basic constituent visual elements of the scene are assembled, the process is cognitively impenetrable, that is it cannot be influenced by beliefs or expectations (Bruce *et al.*, 1994). This raises the question of a means for connecting, objectively, the visually perceived environment with the aesthetic response.

One theory of the structure of memory suggests that it is a pattern of activity within the brain. A particular memory is retrieved by a cue triggering a pattern within the entire range of connections that make up the memory potential. Thus, from a small part of a known pattern the rest can be reconstructed. This forms part of the "connectionist model" (Bruce *et al.*, 1994) and obviates the need for a catalogue of symbols or images that have to be searched through one by one.

The context of such patterns, such as the relationships of objects to each other and to their settings, also becomes embedded in the activity pattern within the brain. Thus, when we see a unified scene, where all the parts relate to the whole, it is possible that we can more easily make a pattern of connections, as these are more readily activated in the brain. The result is that some patterns fit together better and might help to explain why some patterns offer greater aesthetic appreciation than others.

From the basic process of visual perception and recognition of objects or patterns, we can conceive they may have some relationship with the cultural context in which they and the observer exist. It has been established from the discussion so far that we look at a scene with a purpose in mind: to enjoy a natural prospect, to find a route through it, to evaluate it as a good place to live, for example. Thus, some people will see some objects and patterns and not others; yet all may believe that they perceive the reality of the scene in objective terms.

As well as perception of the scene as it is, we can start to imagine how it might be changed in order to serve our affordances better. This expands the process of pattern recognition, firstly to attaching values to some parts of the scene and thence to seeking creative possibilities, projecting from the scene "as it exists" to "what might be".

One primary use of perception is to create models that explain to us the underlying order in what we see. Hence, we may use pattern recognition to abstract from the sensory data a simplified and powerful image (symbol, metaphor) that somehow is the distillate of everything we perceive. Images can describe or visualize abstract concepts, such as liberty or justice (woman with torch, woman with scales). It is possible that we also tend to reduce the landscape into an impression that satisfies our need for generalities, which can be related to many of the preconceived ideas we possess, about the way the world is or how we think it ought to be. In order to be able to apply a general theory, we have to abstract from the circumstances of a case, to recognize the *particular* pattern as belonging to a *class* of patterns. This link between perception and symbolism may have particular relevance to aspects of forestry, such as patterns of logging which may be classified by some people as belonging to a type of "bad landscape" pattern. This has implications for our understanding of forestry aesthetics and ethics.

3   The Aesthetics of the Landscape

In this section, some aspects of aesthetic philosophy are reviewed and linked to the processes of perception outlined earlier. From this, some concepts are developed that have implications for the way we understand forest aesthetics and forest design, particularly in the context of ecological sustainability.

The subject of aesthetics is frequently misunderstood. It is often taken to refer to the sensory pleasure gained from contemplating works of art in a detached or academic way. It has associations with an elite culture, accessible only to those who have learned to appreciate it. The aesthetics of landscape has often been linked with carefully designed pleasure grounds or artfully constructed vistas, where the scene recreates a landscape painting, thus relating the perceptual sensation once more to fine art. The beauty of nature is also part of the traditional realm of aesthetics; in recent times this has become associated with the myth of a pure world, untouched by human activity. Wilderness is its most extreme manifestation. As a result of the association of landscape aesthetics with fine art and natural beauty, the landscape has tended to be treated as a separate object. This has segregated it from everyday life and it is often treated as an optional extra or irritating constraint to the utilitarian demands made of the same scene. This has applied to forestry management as with everything else, and has arguably led to a number of the contentious issues surrounding the future of forests (Bell, 1999).

In contrast, this chapter explores a different set of connections between the basic mechanisms of our sensory perception and the environment, described by Berleant (1992), as "the field of everyday human action", which also has an aesthetic dimension.

3.1 The Nature of Aesthetics

The term aesthetic comes from the Classical Greek words *aesthanesthai*, to perceive, and *aistheta*, things perceived. Thus, perception is at the heart of the way we understand our environment and in this section we will see that it is also central to our sense of beauty and the pleasure we obtain from our environment. Aesthetics is basic to human nature.

We need to express the meaning of aesthetics in perceptual terms and relate this in a practical way to our everyday experience of the environment around us. In this context, the term aesthetics refers to the perceptual and sensuous features of the landscape we experience. All the senses are at work, although, for practical design purposes, sight may receive more attention.

3.2 Environment and Landscape

Berleant (1992) has examined the meaning of the term environment. He suggests that we are used to experiencing the dichotomy of ourselves as separate entities from the world around us, *the environment*. Thus, the concept of the word environment carries certain philosophical assumptions about our notions of self, our experiences and the world around us (Berleant, 1992). Even when we think of environment more generally as "our surroundings", we tend to give it a concept or image, which converts it into some kind of object that cannot easily be grasped or classified.

Alternatively, Berleant presents a more integrated approach in which we and our surroundings are united to form an interdependent whole. We are a part of our surroundings and our perceptions, as discussed above, are the means by which, through our behaviour and actions, we exchange and feedback information to it, resulting in the dynamic, cultural landscape. We should no longer refer to "the environment", because the inclusion of the definite article implies separation of it from ourselves.

Thus, for Berleant, environment is everything: nature, culture and ourselves in an interconnected system. Consequently, all participants, all processes and all human activities that make up the world have to be taken seriously. Extending the meaning of environment also enlarges the domain of aesthetics. We experience the aesthetic dimension of all environments, from the apparently unspoilt, natural beauty of a national park to the commercial landscape of shopping malls and derelict industry. Everything and every experience yields some kind of aesthetic dimension; all impinge on our sensory perception and as that perception is immediate, our reactions can also be very quick.

However, it is also useful to refine the definition of *landscape*, because of the many connotations it holds; it also has a utility and a flexibility as a concept that is easier to understand as an everyday reality for many people than *environment*. In the sense that environment includes everything, the landscape becomes that part of environment which is the field of our present actions. Landscape is the part of environment that we can engage with at a given time.

Landscape can also be understood in terms of the scale of area over which our actions take place. We experience this wide arena composed of different elements and structures during our daily lives. This links well with the Gibsonian approach to perception and affordance examined earlier in the paper. In many ways, therefore, landscape is that part of the environment that is the human habitat, perceived and understood by us through the medium of our perceptions. We cannot escape it and it must not be treated as an optional extra, nor solely as a place to visit on special occasions. It has a powerful effect on the quality of our lives and should not be left to experts or market forces. This awareness has many implications for planners, designers and managers of many types of landscape. There is a direct link here, between the patterns and processes that make up the land, our perceptions of them, and our constant aesthetic engagement that converts the physical dimension of land into the perceptual one of landscape.

3.3 What is an Aesthetic Experience?

Having defined what is meant by the terms 'aesthetic', 'environment' and 'landscape', it is possible to examine what is or is not an aesthetic experience. Foster (1991) has explored in depth the components of an aesthetic experience. Throughout our lives we constantly use our senses to perceive the landscape of which we are a part. Some aspects of this perception yield distinct sensations. For example, we may find beauty in some scenes and ugliness in others. Some places seem to be composed of parts that fit well together and are unified, whilst other places are apparently disjointed and difficult to understand. Is this aesthetic experience dependent solely on the perceptual factors involved at the time of perception or are other, non-perceptual factors involved, such as cultural conditions, scientific knowledge or pre-ordained conceptual patterns? Or do we, through perception, gain the initial aesthetic experience and then use non-perceptual factors to put it in context?

These questions involve two schools of thought. These can be called the *integrationist* and the *perceptual*. It has long been recognized that various elements influence our preferences (and hence our judgments in relation to value or quality), such as history, culture, education, experience, social class, personality and the use of the landscape. The question is to what extent and when do these influence the aesthetics experience?

- *The Integrationist View of Aesthetics*
  In this view, the aesthetic experience cannot be separated from and is in fact interdependent with knowledge about the scene being perceived (Rose, 1976; Carlson and Sadler, 1982; Carlson, 1995). Non-perceptual factors are used to validate aesthetic judgement. Carlson (1995) proposes that ecology, for example, should be used to assess the aesthetic quality of the natural landscape. The question of an ecologically based aesthetic will be examined later in this paper, but it is relevant that it is raised by aestheticians as well as ecologists.

- *The Perceptual View of Aesthetics*
  An emphasis on the idea that aesthetic aspects lead to pleasure or displeasure arising from the perceived scene or object at the time of perception is the key feature of the 'perceptual view' (Foster, 1991).

These positive or negative sensations are only dependent on the way the scene or object appears to the observer, not on knowing the history or cultural framework of a scene or object. Perception stops at the perceptual surface and this defines the limits of the initial aesthetic judgment.

Intellectual, non-perceptual factors are added to the scene or object and it is then that history, culture, experience and so on can help in further appreciation and understanding. This is the crucial point where an understanding and appreciation of the origin and dynamics of patterns in the landscape enters the equation. It is the raised awareness of the structure and processes of the perceived landscape that illuminates the relationship between the aesthetic response and the changing scene, especially in terms of natural landscapes such as forests.

There is also a more specific quality arising from the interaction of sensory elements with social aspects. This is the experience of a sense of place; this can often be a powerful aesthetic dimension, but may only be manifest with some degree of knowledge. This knowledge and the social aspects enhance the aesthetic awareness from the initial sensory apprehension into something with meaning as well as value.

Foster (1991) considers that the interdependent viewpoint of the integrationists is weak, because their focus on conceptual and non-perceptual issues diminishes the impact of the appearance. The appearance, especially the aesthetic component of landscape, is attached to a physical entity of which we are a part (Berleant's self-environment continuum). There is also the specificity of aesthetics as a concept, which has to be separated from interdependence with other social and cultural factors. If we can identify these other factors, because they are more tangible, we should be able to do the same with aesthetics as a field of endeavour in its own right. Then aesthetics can interact with other factors in a much more influential way.

A further weakness of the interdependent view of the integrationists is that if our aesthetic response is conditioned by factors such as knowledge and familiarity, we should find new scenes difficult to appreciate. However, new scenes of which we have no prior knowledge or cultural reference are also stimulating and contain an aesthetic dimension which can only be based on the initial perceptual response.

Nevertheless, the perceptual approach is not straightforward. According to Berleant (1992), although comprehending the landscape is a necessary precondition for a subsequent multi-sensory, aesthetic experience, it is not only comprehension that prompts aesthetic awareness. The perceptual realm is also engaged in ways that enhance and intensify aesthetic awareness. In the appreciation of landscape, there is deliberate attention paid to its component features within the overall setting of our engagement. This is compatible with the Gestalt notions discussed above about making sense of the perceived scene, but extends them to include judgement.

As discussed in the exploration of perception, we do not comprehend our surroundings in a passive way, but actively search out and direct our entire sensory apparatus so that it accepts and interprets sensuous qualities. It follows that this must be a continuous process when we are awake; thus, the landscape can be regarded as a constant source of aesthetic experience. By the same argument, our whole environment is a perceptual system with which we are in constant engagement. However, it cannot be the case that our perception supplies only an aesthetic experience, because we are also seeking many other types of information through our senses. Thus the aesthetic dimension is one element of the total experience and presumably varies in its intensity depending on the situation and context.

Disregard of Aesthetics

If we accept that the concept of constant aesthetic engagement as a part of our total perceptual experience is correct, it is of concern that it has been so often overlooked and frequently ignored. For large areas of activity in the landscape, aesthetics is treated merely as an overlay to our everyday activities. Aesthetics is frequently regarded as an optional extra in commercial development and experiencing the aesthetics of a landscape, particularly its natural components, is relegated to a special activity to be consumed as an alternative to other pastimes or theme parks. Set-piece vistas and specially laid out scenic routes separate the viewer from the environment whilst the experience is filtered by the camera or video lens or car windows.

Without perhaps realizing it, there is a tendency to expect and accept a dull mediocrity in our everyday surroundings. We have to make a special effort to see natural beauty and refresh ourselves. Many people do not or cannot make the effort and so are progressively alienated from their environments (Sircello, 1975). Aesthetic sensitivity becomes moribund and a cycle of lack of care establishes itself. This need not be so if aesthetics was recognized as a vital aspect of a full life and treated with the same importance as the profit motive.

According to the perceptual view, the aesthetics of landscape is open to everyone and requires no prior knowledge of it. Our subsequent assessment is always affected by social and cultural concepts, which can focus on distinct qualities and help set the aesthetic experience in a personal and cultural context. Thus, we gain an immediate aesthetic experience when we perceive the landscape which deepens when we have knowledge of its origins and can appreciate it for its natural or cultural characteristics.

3.4 Pleasure as a Part of the Aesthetic Experience

If part of all sensory experiences is the aesthetic, it is important to know to what extent this experience is enriching to us. If we wish to protect, manage or create landscapes that enrich our lives, we need to recognize that enrichment occurs to a large degree by providing pleasure. Sensory immersion in a landscape can yield different degrees of pleasure or displeasure depending on a number of factors, both perceptual and non-perceptual as discussed above. Strong emotions may be aroused when we experience particularly beautiful environments. For a short while, we may be stimulated to forget personal concerns or become transfixed by the vastness of a scene that takes our breath away. The converse occurs in the face of devastated landscapes, of slums and polluted, derelict industrialized and degraded areas. In surroundings of sheer ugliness, many people learn to ignore aesthetic sensibilities and this may foster a dulled response or emotional inertia to the spread of more ugliness.

The range from the most beautiful landscapes to the most ugly includes those where the scene invokes a neutral response. In many ways this is the most meaningless landscape of all. When faced with ugliness, we may feel impelled to do something about it, but the spiritless, bland landscapes tend to leave us enervated and passionless. These tend to be most numerous in light industrial and commercial suburbs, industrial agricultural areas and transport corridors. We can be demoralized by endless repetitions of unimposing houses, green spaces, warehouses or factories, featureless crops, monotonous plantations and dull roadsides. If we consider that aesthetic pleasure is important, how do we define it and can we have any control over it?

Beauty and the Sublime

At the high quality end of the aesthetic spectrum, lie the twin considerations of beauty and the sublime in landscape. In the 18th century, much intellectual energy was expended developing these concepts by Burke, among others, describing their attributes and differences (Burke, 1958). As concepts, they significantly influenced the direction of landscape gardening and design, especially the picturesque tradition, and became fundamental aspects of landscape appreciation up to present times. They were also fundamental criteria in the founding of national parks in many countries.

Kant (1981) spoke both of free beauty, where the judgment that a scene is beautiful does not depend on concepts determining the nature of that judgment, and dependent beauty, which does depend on such concepts. This echoes the arguments over the perceptual or interdependent nature of the aesthetic experience discussed earlier, and the affordance view of perception.

Schopenhauer (1969) drew a distinction between the presence or absence of "will", which can be interpreted as our self-conscious interest in objects. Any notion of practical interest in what we perceive "falsifies" the way in which we perceive it. However, if during the act of perception of beauty, our mental or physical state requires no control or personal investment of interest in a scene, we are in the right frame of mind to achieve a pure aesthetic experience (Schopenhauer, 1969). This leads to a state where the observer and the scene or landscape being perceived are no longer separated and a respite from the usual world of the will or choice is achieved. In many ways, this bears out Berleant's insistence on the lack of a dualism between humans and environment in the aesthetic realm (Berleant, 1992): during a free aesthetic experience the observer and world become one. Also, for Schopenhauer, there is a great sense of beauty to be obtained if the scene expresses what he calls the "idea", which we can interpret as its essence, or perhaps its spirit of place (*Genius loci*).

Schopenhauer (1969) sees beauty in a landscape when a diversity of objects are found together, when they are clearly separated and distinct, but at the same time "exhibit themselves in a fitting association and succession." This definition is applied primarily to the natural world, but there need be no conflict extending it to many aspects of cultural or designed landscapes. This concept also accords with the three objectives of good visual design expanded by Bell (1993), namely diversity, unity (the fitting association and succession), and the expression of spirit of place, or *Genius loci*. This demonstrates the developing link between the subjective perception of beauty and a rational description of its characteristics, which can be universally experienced.

A sublime experience occurs when our senses are swamped by the magnitude of a landscape that is difficult to comprehend and suggests limitlessness. The imagination and capacity for judgment are also overwhelmed by this impression, as when we try to comprehend the infinity of the universe. This is often one of the initial feelings experienced by many people on first seeing a view of massive mountains or over an apparently limitless desert or forest. We tend to feel very small, humble and helpless when faced with the scale of these scenes or the awesome power of volcanoes, glaciers or hurricanes. The feeling of *potential*, but not *actual*, danger gives the experience an extra sharpness, such as might be felt when looking over the parapet into the depths of the Grand Canyon (Foster, 1991).

Our response to beauty can occur quite spontaneously when the right ingredients are present (supporting the perceptual view). The scene can be mainly natural or be cultural or designed. Order and diversity are necessary: these are found in many places, but not always with the key ingredient of a relationship to spirit of place. Many of the uninteresting landscapes, that are not recognizably ugly, have order in the sense of repetition, but lack diversity and spirit of place. As recognized by Kaplan (1988), there is frequently a wide consensus about what are mainly beautiful landscapes; the point that may be debated is where they cease to be beautiful and descend into ordinariness or blandness.

A question arises as to whether the deeper understanding of natural processes and patterns diminishes the sense of the sublime. This may be so, if we adhere to the integrationist approach, as many of the chances to perceive beauty and to experience the sublime would be lost, because of the need for prior knowledge. This would not be so if we prefer the perceptual approach.

The two theories of beauty and the sublime remain rooted in the notion that we are external observers, using the *aesthetic senses* of sight and hearing. Is it possible to move on from this and to incorporate the multi-sensory engagement of the self-environment continuum? There is another, overlooked set of theories that may help achieve a fuller integration of approaches: that of Whitehead.

The Aesthetic Theory of Alfred North Whitehead

Gunter has developed the links between a metaphysical theory of environment and aesthetics postulated by the philosopher and mathematician Alfred North Whitehead (Whitehead, 1960; Gunter, 1996). This theory presents some useful ideas that may have been overlooked, because Whitehead often uses complicated and ambiguous terms which make his work difficult to understand.

Metaphysics is a part of philosophy that investigates the essence and relationships of living things (being) and of reasoning (knowing). Whitehead's metaphysics and his philosophy of nature extend both terms into a wider application that, in some cases, includes non-living things. Whitehead refers to living things as organisms, and the way they observe and reason about their environment as prehensions (as in apprehend or comprehend).

All aspects of an organism's prehensions merge into what Whitehead calls concrescences, meaning coherence, or coming together. Thus organisms, like ourselves, can be said to exist as a fusion of their character and the sum of all their prehensions, or interactions with their environment and other organisms.

Whitehead then develops this metaphysical concept into an aesthetic theory. He sees us existing in our environment as living organisms among other organisms, undertaking prehensions. The sense of beauty is gained as a result of positive prehensions and ugliness due to negative ones. This leads to his recognition of two forms of beauty, a *major* and a *minor*. The absence of discord or painful clash amongst the prehensions (and presumably the absence of a negative concrescence) or the absence of "mutual inhibition" gives the minor form. We may liken this to the basic level of unity in a landscape, of a lower end of attractiveness where there is no obvious discordance or imbalance, but where our senses are not significantly aroused.

The major form requires the existence of the minor form before it can occur, and builds on this. It involves one or more contrasts between the factors of perception, which provoke an intensity of feeling. This increases the intensities of the basic component feelings, so there is a positive feedback that further increases the overall intensity of the experience. There are two characteristics of the landscape that provoke these intensive feelings. The first Whitehead calls *massiveness* or the presence of a variety of detail with effective contrast. Massiveness includes diversity or complexity, but does so at a range of scales. The second characteristic is called *intensity proper*, meaning comparative magnitude (or scale), without reference to the variety of detail that gives massiveness. Intensity proper could be linked with the magnitude that induces a sublime experience. Where massiveness and intensity proper occur together, it is conceivable that a form of order or unity would exist.

The scenes or environments that we find the most beautiful or attractive should, according to Whitehead's theory, contain massiveness and intensity proper. Before examining some examples, we need to assess how the idea of prehension and aesthetics is invoked. When we seek to comprehend a landscape, we may use our eyes to look at it, some parts distant, others close up; we also feel the solidity of the earth beneath our feet and can touch nearby surfaces. As we do this we are taking aspects from this landscape and incorporating them, using the mechanisms of perception, into part of ourselves both mentally and physically. We thus also gain affordances, in Gibsonian terms, by the incorporation of these prehensions into an organism (ourselves) and we maintain our place in the self-environment continuum as defined by Berleant.

The minor form of beauty, "the absence of painful clash, the absence of vulgarity" ensures that the component parts of our perceptual experience do not inhibit one another and prevent the feedback of intensity.

A forest example shows the application of these properties: in a plantation of pine trees, all the trees are the same age, planted in rows, without understorey, tall with cylindrical trunks. Here is comparative scale, but little or no qualitative variety; thus it has intensity proper but lacks massiveness. In contrast, a natural or semi-natural pine forest has a range of ages and sizes of trees of varying species, together with undergrowth and wildlife. Each tree has a different form, some twisted, some branching, some straight. This kind of forest has massiveness as well as intensity proper and is a definite possessor of the major form of beauty (Gunter, 1996).

The culmination of massiveness and intensity proper in the greatest degree is referred to by Whitehead as *strength*; in some ways it is similar to the strength of character that distinguishes the most attractive landscapes, possessing a powerful sense of place, from ordinary landscapes.

Whitehead's scenario of beauty is invoked by positive contrasts, so a problem might arise if the otherwise beautiful natural scene has discordant natural elements within it such as trees blown down or burnt areas. However, these can only be considered discordant, if obsolescent knowledge of nature's processes is applied. Beauty arises from regeneration, and massiveness can only develop from the processes that change the simple pioneer stands into the complex structural and spatial pattern of the fully developed forest. Even if these factors are seen as negative, they can provide the extra sense of raw, vibrant life that translates the merely pretty into a form of mature beauty, whose strength arises in part from the reality of nature.

From the initial consideration of what constitutes an aesthetic, Whitehead's theory allows a definition, albeit in metaphysical terms, of what characteristics can determine beauty and the sublime. In the discussion on the importance of post-perceptual knowledge, we saw that the aesthetic experience can be deepened if we know more about the landscape. In the appreciation of nature it would seem logical that there should be some direct links between aesthetics and ecology, as suggested in the example of the natural pine forest described earlier. This will be taken up later in the chapter.

The Individual Versus Universal Appreciation

The cliche "beauty is in the eye of the beholder" is, as many cliches are, a self-evident truth. For many, it means that everyone perceives a different quality in a landscape and that some find a thing beautiful that others find ugly. This may be so if the perception and appreciation involves the will and an interest in the scene under perception. A logger may not find a logged mountainside ugly, whereas a committed environmentalist may well do so. This also involves the role of knowledge in the aesthetic experience.

However, we each possess the same sensory faculties and, apart from those who are impaired, have the same access to the perceptual surfaces, sounds, smells, tastes and kinaesthetic responses. We all have brains with similar mechanisms able to process and interpret the basic perception and to find order, pattern, structure and sense in the world. With a few exceptions, designers have been able to create landscapes that the majority of people find attractive or beautiful; it tends to be a minority who claim to see beauty in some universally decried scene.

Thus, we can postulate there ought to be a high degree of universality in the acceptance of a sense of beauty or sublimity when people are presented with certain landscapes. Personal preferences, cultural overlays and practical involvement if applied post-perception, are what should be responsible for producing the nuances of moral or ethical standpoints that are commonly encountered. It is the role of preference research to uncover some of these factors and to try to disentangle those that are universal from those that are dependent on culture, education, experience, ethical position and so on. It is also necessary to try to ascertain what factors pertain to the characteristics of the landscapes and those that pertain to people's perceptual processes. Luckily, a good deal has been done over the years, but does it get us any closer to the answers?

3.5 Measuring Aesthetic Appreciation and Preferences for Landscape

Many landscape preferences studies have followed the scenic beauty estimation method (Daniel and Boster, 1976) or variations of it, using ratings of photographs of different landscapes. Many such studies of forests have been carried out and it is possible to find that some generally held preferences for certain forest landscapes exist across a range of examples. These preferences can be summarized in terms of the landscape elements present, such as large, widely spaced trees. However, these findings are so general that they are of little help to practitioners.

Further studies attempt to predict what people's preferences will be in relation to the content of different scenes (Thayer *et al.*, 1976). These are of little practical use in terms of landscape management because of the lack of mechanisms for integrating visual or aesthetic management into the wider field of forest management, let alone any link with forest ecology and naturalness. They also assume that the aesthetic experience is an isolated one with no sense of other factors entering the equation.

Some preference studies have explored the links between landscapes that not only provide a pleasurable aesthetic experience of beauty, but also to fulfil a function (Lee, 1991). The question may be asked: in what terms do I prefer this scene over that one? As a place to live? As a place for a picnic? As a place to see wildlife? This brings more than perceptual values into play and uses the landscape, not only as the setting, but also as an integral part of the overall experience. This approach was psychophysical, linking the cognitive aspects of the perceivers with some of the characteristics of the landscapes perceived. This did not achieve very high correlation between these factors, and any common preferences were very general.

While it is understandable for planners to want to know what effects their landscape changes are likely to have on the people who experience them, psychologists are probably more interested in the cognitive attributes that are present in people when they experience a landscape that gives them pleasure. In other words, too much concentration on the characteristics of the landscape at the expense of understanding the mechanisms of perception and aesthetics is likely to miss the point. In fact, for planners to concentrate more on cognitive factors may prove to be a good deal more useful for practical purposes. For example, Kaplan (1988) has shown that cognitive features common to nearly all highly rated landscapes can be summarized as:

Coherence: the ability to see and comprehend the pattern inherent in a scene (the opposite of chaos)

Complexity: the range of different elements in an object or scene that provides sensory stimulation

Mystery: the aspects of a scene that cannot be comprehended all at once.

We will return to these factors later, because they are also very similar to ones derived from other approaches such as those of Whitehead.

However, it remains the case that practitioners need some guidance so that they can manipulate the landscape in ways that they hope will produce positive responses in the users or viewers. By looking at the subject in a different way, this may be attainable.

3.6 Characteristics and Qualities in the Landscape

It is important to remember that our perception tends to be of the world as a total environment (obtaining a sense of coherence), rather than a mass of single isolated objects. To achieve this requires the ability to identify the link between the characteristics of the whole in relation to the sum of the characteristics of the parts (which yields complexity).

Schopenhauer (1969) linked beauty to the connection between the parts as being much more than those of the parts themselves. Whitehead defined characteristics such as massiveness, intensity proper and strength in terms of a minor and major form of beauty (Whitehead, 1960; Gunter, 1996). Designers, such as landscape architects, seek to synthesize the parts into a unified whole and use a range of techniques to achieve this (Bell, 1993). A link is now needed between the aesthetic description of environmental characteristics, both physical and perceptual, the qualities they yield (subjective-value laden, aesthetic) and the application of that description to practical design. It can be found in the intertwining of the philosophical strands developed by aestheticians with the empirically derived principles employed by designers.

From the exploration of perception, it has been established that environmental aesthetics is not simply the successive perception of different objects, but how we comprehend the pattern of the entire scene as a coherent or unified whole. Thus, the characteristics inherent in the scene combine to produce the aesthetic qualities perceived by the observer, such as coherence, complexity, or mystery.

Emotions generated by landscapes such as agoraphobia, claustrophobia, serenity, calmness and so on, are the responses associated with subjective, qualitative characteristics such as openness, wildness, bleakness, intimacy, etc. Each is achieved in part due to the interactions of various characteristics. The emotions felt are personal, although many may also be universal. For example, openness is due to a lack of enclosure, a large scale in proportion to the space and the textures of the surfaces, whilst intimacy is the product of strong enclosure, small scale (in relation to our size) and small size of the constituent parts.

Thus, we can say that a landscape is constructed from a number of constituent parts (each possessing describable characteristics, as does the whole scene), where the parts act in combination. The sense of place or the emotions provoked by the landscape are qualities experienced by the observer, but depend partly on the characteristics possessed by the landscape and partly on the non-perceptual aspects possessed by the observer. Designers can usually only control one set of variables, that of the landscape, so should only try to work as far as they can towards meeting preferences that are as universal as possible within the population who are the users or observers. However, what happens when the landscape changes are unable to meet the expressed preferences of these users because of other factors such as ecology? Is it possible to change perceptions and preferences also?

3.7 Ecology and Aesthetics

Leopold (1949) attempted to develop an ecological aesthetic that could transform the way people change the landscape. This was based on his land ethic, where he claimed that "A thing is right when it tends to preserve the integrity, stability and beauty of the biotic community. It is wrong when it tends otherwise" (Leopold, 1949, p. 262). Gobster (1995), writing about Leopold's ecological aesthetic theory, believes that this change in focus broadens our ideas of the aesthetic experience, expanding it from the essentially visual, and immediate pleasure gaining process, to one that uses all our senses and our intellect. This includes aspects of the non-perceptual in aesthetics, which was discussed earlier. This approach challenges traditional views of landscape aesthetics, based on the scenic aesthetics of contemplation. The question raised is: can we enjoy a beautiful landscape all the more if we also know it is ecologically healthy? Clearly some knowledge of the landscape in question is necessary, but this is not absolutely essential for the perception of the initial aesthetic experience, which may therefore be negative.

Since Leopold's theory comes from an ecologist, not an aesthetician, it throws new light on aesthetics, for as Gobster (1995) argues, the landscape we observe can become more comprehensible as our focus widens to include more complex layers of ecology. The dramatic scene continues to give pleasure either in the Kantian mode or in terms of complete sensory engagement; nevertheless, the greater degree of exploration and appreciation of the dynamics of natural change leads ultimately to a deeper satisfaction and elucidation of the spirit of the place. This is exemplified by indicators of aesthetic/ecological health, such as the range of wildlife species present.

Leopold also said, "Our ability to perceive quality in nature begins, as in art, with the pretty. It expands through successive stages of the beautiful to values yet uncaptured by language" (Leopold, 1949, p. 102). This is an essential link with Whitehead. Aesthetics begins with pleasant but unspectacular (the minor form of beauty) and leads to the major form of beauty and the strengths of the ecologically healthy landscape, expressed in its massiveness and intensity proper.

This connection with the natural and cultural patterns and processes provides a most rewarding vein for aesthetic exploration. The link completes the circle from the structure of the landscape to our basic perceptions of it; if we are able to make sense and orientate ourselves in the landscape, whilst appreciating the process involved, this can lead to the highest aesthetic experience. This can also yield fresh experiences every time we engage with the landscape.

Brennan (1988) emphasized that we are part of the natural world as well as the cultural one. He developed a notion of *ecological humanism* (Brennan, 1988), underlining his belief that if we live nearer the ecological reality of the world, our lives will be more rewarding and fulfilled. What we are comes, partly, from where we are. Thus, the identity of place strongly influences our views as individuals and those of the communities in which we live. In this case there should be a strong link between the ecological health of the environment as a whole and our own sense of well being. An understanding of the ecological processes at work in our own neighbourhoods should deepen our desire to achieve a better landscape and so help to overcome the disregard of aesthetics noted earlier.

At its best, the contemplative mode that can be engendered when we engage with nature or wild surroundings, and the sense of beauty or the sublime that helps to leave our cares behind, is more than a detached Kantian contemplation. It can be the fullest expression of complete engagement, responding to a higher level of environmental beauty.

Engagement with the natural world can put many of us in harmony with the cycles and rhythms of life, such as the changing seasons, the tides or animal and plant life cycles. Sircello (1975), thinks alienation is the great malaise of our time, linking it with a lack of interest in beauty so that we can almost feel as strangers in our own environment.

This alienation, paradoxically, occurs at a time when natural imagery is a powerful influence on the style of life to which many people aspire. Concern about the environment is very great, but this is frequently misplaced or misdirected through ignorance of the dynamics of the natural world or the realities of rural life; this is due to a lack of contact or engagement with our environment. The car-borne visitation to a national park or other scenic area tends to filter the experience as a detached perception, wholly visual and based on the distant panorama.

Brennan (1988) takes this a step further, asserting that because we are a part of the natural world, we require contact with it to become fulfilled human beings. He links our identification with nature as the first step towards an environmental ethic. In as much as ethics and aesthetics go hand in hand, it is illogical to recognize the direct benefit of our engagement with nature and then ignore its care as much as we do.

Berleant (1992, p. 164), in trying to define how nature differs from art in aesthetic terms, concludes that we need an "aesthetic that bases our appreciative response on the awareness, selection and understanding of the order by which natural forces have produced the objects we admire." This takes us back to the search for coherence, complexity and mystery inherent in beautiful landscapes and manifested in nature, as well as the deeper engagement described here.

This should help us to develop an appropriate aesthetic and ethical position for the environment, perhaps based on Leopold's land ethic. We also need a greater understanding of the ecological and cultural patterns and processes that produce beautiful landscapes.

3.8 Aesthetics and Forest Landscapes

From the discussion so far it should be emerging that forests, as a type of landscape, have particular properties and aesthetic implications. It has been suggested that forests are one of several archetypal landscapes that contain many meanings for us, of which the aesthetic is one (Schama, 1995).

Firstly, in terms of perception, there are distinctions to be made between internal and external views and the uses of the senses. The perception of the external, scenic view is primarily visual, most open to appreciation in the contemplative mode proposed by Kant (1981), and the internal view from wandering in the forest, most open to the multi-sensory aesthetic of engagement proposed by Berleant (1992).

This perception also relates directly to the approach advocated by J.J. Gibson (1979) and his affordance theory. When we visit or experience a forest we gain affordances from it. These may vary, depending whether we work or play there, have a utilitarian requirement or not. When we compare the differences between Schopenhauer's unself-conscious and self-conscious attitudes (the presence of the "will" at the time of perception), we can also see that different affordances produce different modes of aesthetic awareness and experience. Thus, visitors, residents or managers of forests may, while all the time undergoing constant aesthetic engagement with the forest, gain different things from it and be prepared to desire beauty or the sublime to different degrees.

In perception we are also trying to make sense of the world, to understand it and our place in relation to it. It may be that if we place more importance in the external view, we are able to use the awareness of Gestalt theory on how and why some patterns of forest layout work better and appear more attractive than others. The development of design principles is easier and their implementation more straightforward. When within a forest, our senses may struggle more to cope with the multitude of sensory stimuli.

On the one hand, being immersed in a landscape, such as a forest, may enrich our aesthetic experience because all the Whiteheadian characteristics of massiveness, intensity proper and strength come into play, as we exercise our prehensions. If the forest possesses the major form of beauty then we are likely to have a pleasurable experience. However, at the same time, we might find the forest confusing and we may be fearful of getting lost. On the other hand, from a utilitarian perspective, we may consider that unruly nature, and the effects of natural processes are offensive to our ideas in relation to our affordances for wood production. To tidy it up prevents us from getting lost, because the Gestalt laws are easier to operate and we are better able to perceive the patterns. However, this reduces the sense of massiveness that we saw in the plantation example in our discussion of Whitehead.

There are aesthetic complications in terms of modern forestry management, since its history has been one of simplifying forest structure and pattern in the interests of more efficient wood production. This reflects the separation of people from the environment. It also runs counter, in many ways, to the land ethic and ecological aesthetic of Leopold. In addition, the history of fire suppression, pest control and logging practices deny the place of natural processes in the landscape. The irony here is that the integrationist position should make it impossible for anyone to find any type of managed forest attractive, because management takes away the purity of the forest's naturalness. In fact, well designed forests can be found attractive as long as they produce a sense of coherence, complexity and mystery, whether they are managed or not, purely natural or not.

Thus, we can see that there are tensions in the aesthetic of the forest. These occur between the contemplative mode of Kant and the engaged mode of Berleant; between different Gibsonian affordances and Schopenhauer's will; between the desire for a readable landscape with simple Gestalt properties and the richness of one with Whiteheadian characteristics; between the orderly, human-controlled world and the Leopoldian ecological aesthetic and, finally between the perceptual nature of aesthetics and the integration of knowledge. We can resolve these tensions by remembering the qualities of coherence, complexity and mystery and their role in the identification of landscapes preferred by many people arising out of perception research. We can also broaden the scope of design to balance the competing tensions. A major way of achieving this will be through the links between aesthetics and ecology, and the need to integrate processes of change into the forest based on Leopold's land ethic. Participation by communities working with experts will also overcome some of the problems associated with changing landscapes, with different affordances and with alienation from the environment.

4 Conclusions

Forests occupy an important place in our culture. They are one of the few types that enclose us and separate us from the outside world, that gently stimulate all our senses in their perception and, when we suspend our will, naturally calm us. They provide beauty and elements of the sublime, but to do so they need to invoke a sense of coherence, complexity and mystery. Preference studies can tell us a great deal about what types of landscape people prefer, but to be able to answer the deeper questions about how and why they prefer them, we need an understanding of the basis of perception and aesthetics. Then we can use design and management to find, with help from the stakeholder communities, the correct balance that provides the richest aesthetic engagement with ecological possibilities and economic benefits.

References

Bell, S. (1993) *Elements of Visual Design in the Landscape.* E. & F.N. Spon, London.

Bell, S. (1999) *Landscape: Pattern, Perception and Process.* E. & F.N. Spon, London.

Berleant, A. (1992) *The Aesthetics of Environment.* Temple University Press, Philadelphia.

Brennan, A. (1988) *Thinking about Nature.* Routledge, London.

Bruce, V., P.R. Green and M.A. Georgeson (1994) *Visual Perception: Physiology, Psychology and Ecology,* 3rd Edition. Psychology Press, Hove.

Burke, E. (1958) *A Philosophical Enquiry into the Origin of Our Ideas of the Sublime and the Beautiful,* J.T. Boulton (ed). Routledge and Kegan Paul, London.

Carlson, A. (1995) Nature, aesthetic appreciation and knowledge. *Journal of Aesthetics and Art Criticism* 53(4): 394-400.

Carlson, A. and B. Sadler (1982) *Environmental Aesthetics: Essays in Interpretation.* University of Victoria.

Daniel, T.C. and R.S. Boster (1976) *Measuring Landscape Aesthetics: The Scenic Beauty Estimation Method.* USDA Forest Service Research Paper RM-167.

Rocky Mountain Forest and Range Experiment Station, Forest Service, Department of Agriculture.

Foster, C.A. (1991) *Aesthetics and the Natural Environment*. Unpublished Ph.D. Thesis. University of Edinburgh.

Gibson, J.J. (1966) *The Senses Considered as Perceptual Systems*. Houghton Mifflin, Boston.

Gibson, J.J. (1979) *The Ecological Approach to Visual Perception*. Houghton Mifflin, New York.

Gobster, P.H. (1995) Aldo Leopold's ecological esthetic: Integrating esthetic and biodiversity values. *Journal of Forestry* 93(2): 6-10.

Gunter, P. (1996) *Organism and Prehension: A Forest Aesthetic*. Unpublished paper presented at Conference on the Aesthetics of the Forest, Punkaharju, Finland.

Kant, I. (1981) (translated by J.H. Bernard) *The Critique of Judgement*. MacMillan, London.

Kaplan, S. (1988) Perception and landscape: Conception and misconception. In: Nasar, J.L. (ed) *Environmental Aesthetics*. Cambridge University Press, Cambridge.

Kohler, W. (1947) *Gestalt Psychology, An Introduction to Modern Concepts in Psychology*. Liveright Publishing Corporation, New York.

Lee, T.R. (1991) *Forests, Woods and People's Preferences*. Unpublished report to the Forestry Commission, Edinburgh.

Leopold, A. (1949) *A Sand County Almanac*. Oxford University Press, New York.

Marr, D. (1982) *Vision: A Computational Investigation into the Human Representation and Processing of Visual Information*. Freeman, San Francisco.

Penning-Rowsell, E. and Lowenthal, D. (eds) (1986) *Landscape Meaning and Values*. George Allen and Unwin, London.

Rose, M.C. (1976) Nature as aesthetic object: An essay in meta-aesthetics. *British Journal of Aesthetics* 16 (1).

Schama, S. (1995) *Landscape and Memory*. Harper Collins, London.

Schopenhauer, A. (1969) (translated by E.F.J. Payne) *The World as Will and Representation*, Dover Publications.

Sircello, G. (1975) *A New Theory of Beauty*. Princeton University Press, Princeton.

Thayer, R.L., R.W. Hodgson, L.D. Gustke and J. Holmes (1976) Validation of a natural landscape preference model as a predictor of perceived landscape beauty in photographs. *Journal of Leisure Research* 8: 292-299.

Whitehead, A.N. (1960) *Adventures of Ideas*. Mentor Books, New York.

Chapter eleven:

Beyond Visual Resource Management: Emerging Theories of an Ecological Aesthetic and Visible Stewardship

Stephen R.J. Sheppard
Collaborative for Advanced Landscape Planning (CALP), Department of Forest Resources Management Department and the Landscape Architecture Programme, University of British Columbia Vancouver, British Columbia

1 Introduction

The dominant set of guiding principles on the aesthetics of forest lands in North America is embodied in classic visual resource management (VRM) programmes, such as the USDA Forest Service Scenery Management System (USDA Forest Service, 1995) and the British Columbia VRM procedures under the Forest Practices Code (BCMoF, 1997a). The largely expert-based procedures have proven to be quite robust as a management tool in the face of sometimes harsh criticism from the forest industry, as well as from some academics. The classic battles in practice have been over the apparent inverse relationship between visual quality and timber supply (Figure 11.1a). In the author's experience, the importance of visual resources has never been solidly accepted by most professional foresters, except at times as a political necessity. Nonetheless, in both the US National Forests (USDA Forest Service, 1981) and in BC (BCMoF, 1998), the application of Visual Quality Objectives (VQOs), which establish levels of allowable landscape change, has accounted for major reductions in otherwise available timber supply. As a result, the mandatory and universal application of VRM systems has been modified or weakened to some extent under both programmes (USDA Forest Service, 1995; BCMoF, 1997b), yet the overall programmes remain in effect and continue to exert considerable influence on forest management.

However, as the movement towards ecosystem management has begun to dominate forestry and as industry shifts towards more sustainable and certifiable forestry (see Chapter 7 in this volume), a more complex debate has developed, one which is at least triangular (balancing timber supply with ecology and aesthetics - see Figure 11.1b) and often multi-dimensional. The visual quality indicators

a) Traditional conflict

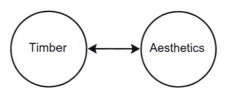

b) Contemporary conflict

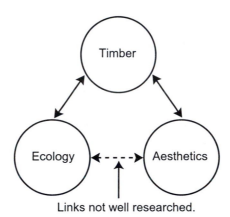

Links not well researched.

Fig. 11.1. Simplified diagram of classic resource conflicts in BC.

typically used in VRM appear rather narrow and less relevant to these more complex issues (Gobster, 1999). A broader question has thus emerged: do visual quality objectives help or hinder the achievement of ecosystem management goals such as biodiversity and ecological health, and vice versa? In short, are scenic landscapes sustainable, and are sustainable landscapes scenic? This question, addressed throughout this volume, can be posed also in terms of whether forest ecologists seeking to promote sustainable ecosystems, are in agreement with the general public, who tend to be least happy about management practices which fail to look attractive (as presented rather simplistically in Figure 11.2). If there is conflict between these management goals, how can they be brought into closer alignment?

This chapter focuses on relationships between aesthetic values and commonly used indicators of sustainability. In this context, I address primarily the dominant ecological component of sustainability, although as will be pointed out, the social and economic dimensions of sustainability cannot be divorced from the larger equation. The chapter explores various aesthetic theories and related forest

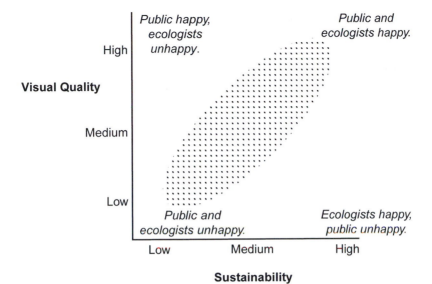

Fig. 11.2. Hypothetical relationships between ecological sustainability and aesthetics: contrasts in public and ecologists' viewpoints (shaded area indicates strong agreement).

management principles that can be used to explain relationships between visual quality and forest sustainability.

These theories are considered under the following categories:

- The conventional *scenic aesthetic*, as it is termed by Gobster (1999), following VRM principles;
- An *ecological aesthetic* which is more in keeping with the *new forestry* and has recently received significant attention (Gobster, 1995; 1999)
- A related but broader aesthetic of *visible stewardship*, proposed in this chapter but drawing on the work of several previous authors and supplementing, rather than conflicting with, the previous two theories.

The following sections discuss each of these aesthetic concepts in turn, touching upon their theoretical basis and supporting evidence, the nature of their relationship to commonly used ecological indicators of sustainability, and problems and limitations (relevant to the relationship with ecological factors) in theory or practice. The implications of the visible stewardship theory, advanced here, are explored further, in terms of possible directions for research and practice.

2 The Scenic Aesthetic

The development of visual resource management (VRM) as a field can be traced to two key events: the requirements of the US National Environmental Policy Act (NEPA) of 1969, which for the first time mandated the consideration of aesthetic effects of Federally funded projects; and the clearcutting controversy of the 1970s on public forest lands in the USA, although analogous controversies had erupted slightly earlier in the UK over planting of coniferous forests on open uplands (Crowe, 1966). Public reaction to the scale and appearance of timber harvesting in the US National Forests led to a mandatory nationwide programme of inventorying and planning to protect important scenic resource values on all forest lands (USDA Forest Service, 1974). Developed by a team of US Forest Service landscape architects drawing on the seminal work of R. B. Litton Jr. (*e.g.* Litton, 1968), the *Visual Management System* established visual quality objectives for all federal forest land, whereby the most scenic and the most visible lands received the highest protection from visual impacts of large scale resource management activities. This programme has since been adopted, adapted, and applied in other parts of the USA and around the world, notably in the public forests of British Columbia' s Crown lands (BCMoF, 1981; Sheppard, 1999)

The VRM movement established aesthetics as a "resource" in its own right, a somewhat tangible value which could be systematically mapped and included in forest management decision-making (Smardon, 1986). Aesthetic values earned an equal seat at the multiple-use table with the other timber and non-timber values represented by various forestry disciplines. VRM was a solid achievement and a successful tool for its time. In effect, it constrained timber production in the most visible areas (USDA Forest Service, 1981), thus fostering increased public acceptability of overall forest management.

Today, 25 years later, VRM continues throughout the US National Forests, albeit in a somewhat modified form (USDA Forest Service, 1995). It still has many detractors within the forest industry, and even within the landscape architecture profession, and it undoubtedly suffered from many methodological shortcomings and considerable unreliability (see Carlson, 1977; Smardon, 1986). However, one could argue that it continues to serve its intended purpose of reducing both visual impacts and public hostility to forestry on public lands.

In fact, in some places it could be argued that VRM has gone beyond its original purpose, in contributing to the public expectation for national forests as naturalistic recreational playgrounds - parks in all but name - rather than multiple-use working forests supplying industrial and other products. At a workshop on scenic values held at Lake Tahoe in 1997, senior US Forest Service management for Region 6 in California announced that clearcutting was no longer an acceptable management tool on their forests (Strain and Sheppard, 1997). More recently, the private Canadian forest company MacMillan Bloedel (now Weyerhaeuser) announced that it, too, would phase out clearcutting on its holdings on the coast of BC. These decisions can be considered in part a response to aesthetic reactions of the public against clearcutting.

2.1 Theoretical Underpinnings of the Scenic Aesthetic

Before we examine the relationship of forest sustainability to aesthetics as defined in the VRM paradigm of the scenic aesthetic, we must review the underlying theories and empirical evidence supporting this perspective on the forested landscape. The VRM approach emphasizes the importance of visual variety in the landscape as an indicator of scenic quality, visibility and viewing distance as determinants of visual sensitivity or concern, and emulation of natural visual patterns (in form, line, colour and texture) as desired characteristics of management activity (USDA Forest Service, 1974).

These tenets of the Visual Management System, as it was first called by the US Forest Service, were derived mainly from the advances in landscape description pioneered by Litton (1968) using the terminology of landscape architecture, coupled with the practical need to make evaluative decisions and choices between management alternatives. Since the system was developed by landscape architects within the US Forest Service, we can assume that they were also influenced by the history of aesthetic appreciation of natural landscapes. This history ranges from the classical movement in art in 18th century Europe with its emphasis on symmetry and clarity of form, to the taste for the *sublime* grandeur of nature and the picturesque of rural scenes, through to the wilderness appreciation movement in North America (Porteous, 1996; Carlson, 1998). However, another key influence (or justification) for the approach were the little publicized findings of Floyd Newby (undated), which emphasized that visitors to the National Forests expect to see a natural-appearing landscape.

There is considerable evidence that the *scenic aesthetic*, as defined by the expert VRM professionals and landscape architects of the US Forest Service, generally fits well with prevailing public perceptions of the landscape. It reflects the fairly consistent results of environmental perception research on forested landscapes, particularly that research focused on scenic beauty (*e.g.* Daniel and Boster, 1976; Zube *et al.*, 1982; Kaplan *et al.*, 1998). In BC, perception research on responses to clearcutting in photographs indicates that in, general, the larger the extent of visual change, the lower the level of public acceptance (BCMoF, 1996). This appears to bear out the approach to VQOs as followed in BC, whereby the more sensitive and scenic the view, the smaller percent alteration of the visible landscape which is allowed. At some level, if one is seeking to avoid conflict over aesthetics, it is very logical to hide the logging activities wherever possible, consistent with the adage "out of sight, out of mind."

The criteria used in identifying scenic quality in classic VRM programmes can be related not only to cultural theories of aesthetics and design as mentioned above, but also to biological or evolutionary theories of aesthetics. These invoke instinctive, conscious or sub-conscious reactions to landscapes as a primary influence on aesthetic response, derived from the need for survival (Appleton, 1975; Orians, 1986): for example, visual variety can be related to the incidence of biologically productive edge effects; panoramic views and vistas can be related to prospect-refuge theory and the desirability of seeing without being seen (Appleton, 1975); and downed wood or heavy screening can be correlated with risk of attack or impediments to escape.

2.2 Theoretical Relationships of the Scenic Aesthetic to Sustainability

The architects of the VRM movement appear to have assumed that aesthetic principles go hand-in-hand with ecological principles. How could any programme which is designed to emulate natural patterns and "look natural" not be ecologically benign? Bradley (1996) provides the corollary to this view, stating that:

> Requirements that result in the retention of vegetation, such as Habitat Conservation Plans, green-tree retention, riparian zone management..., or other aspects of fish and wildlife protection, are all likely to complement efforts to minimize the visual impact of harvesting. (p. 12).

The pursuit of naturalness sets natural conditions as the baseline to be retained (as in the VQO of *Retention*) or modified (as in the VQO of *Maximum Modification*). Visual variety may correspond with biodiversity. The application of landscape design implies a sensitivity to the land which augurs well for ecological sustainability, as postulated by Diaz and Apostol (1992) in the principles of *ecological design* (see also Chapter 10 in this volume). According to these practitioners, such an approach seems to work in practice, to fit both landscape "needs" and public desires.

The *Scenery Management System* (USDA Forest Service, 1995) also attempts to integrate the scenic aesthetic with ecosystem management principles, though more by explicitly allowing trade-offs than by aligning fundamentals of the two sets of values.

2.3 Problems and Limitations of the Scenic Aesthetic

As Gobster (1999) and Nassauer (1997) point out, there are clearly situations where that which looks good is not necessarily good for ecosystem health or sustainability. The goals of ecological management and aesthetic perceptions can conflict, for example with the occurrence of fire necessary to rejuvenate the landscape, high levels of dead and downed wood for improved forest health and biodiversity (Figure 11.3 and Plate 7), and the reduction in forest fragmentation by increased clearing sizes with minimum edge-to-area ratios. Rountree (1998) has documented the adverse impact on forest health of measures such as fire suppression, intended to protect scenic landscapes in the Lake Tahoe Basin. Klenner (1999, pers. comm.) has described the problems of trying to gain public approval for large aggregated harvesting units which are intended to emulate natural disturbance patterns. Conflicts also transcend the forestry practices themselves, and can arise from the large-scale spatial allocation of forest harvesting: in many

Fig, 11.3. Heavy concentrations of downed wood usually attract adverse aesthetic ratings in perception experiments. Source:BCMoF (see Plate 7).

areas of BC, preservation of the *front country* most visible from highway corridors and communities has led inevitably to extensive exploitation of the more pristine old growth and intact stands in the less visible *back-country* (Picard and Sheppard, 2000).

Clearly therefore, it is possible to identify various situations which occur outside the zone of compatibility between the scenic aesthetic and ecological sustainability, as shown in Figure 11.4. A more systematic categorization of conflicts and compatibilities between ecology and aesthetics in forest harvesting is shown in Table 11.1, which represents an initial attempt to compare key indicators of ecological sustainability (as exemplified by "new forestry") and scenic aesthetics, across landscape scales.

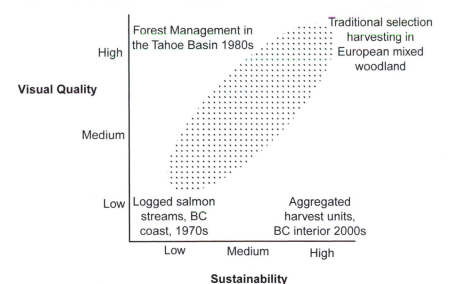

Sustainability

Fig. 11.4. Examples of relationships between sustainability and aesthetics (shaded area indicates strong agreement).

Table 11.1. Comparison of selected indicators of sustainability ("new forestry" ideals) and scenic aesthetic ideals for timber harvesting at different landscape scales.

Landscape Scale	New Forestry Ideals	Scenic Aesthetic Ideals
Tree/stand (foreground views)	*- Downed timber/debris* - Snags - Natural regeneration - No erosion	*- Clean forest floor or clearing (few stumps)* *- Grassy/mown strips/ edges* - Large trees at open spacing - No erosion - Signs on forest restoration *- No fire-scars/ scorched earth*
Stand/cutblock (middleground views)	- Balanced edge to interior ratio - Minimized windthrow - Established succession/ regeneration - Riparian leave strips	- Irregular cutblocks - Feathered cutblock edges *- Cutblocks in lower, flatter locations* - Green tree retention *- No road scars* - Rapid green-up
Landscape/watershed (middleground/ background views)	- Unfragmented ecosystems - Balanced patch size distribution - Intact riparian corridors *- Occasional large scale disturbances* *- Harvesting in already disturbed (second growth) front-country areas*	- Irregular cutblock distribution emulating natural patterns - No extensive road scars *- Harvesting in unseen (back-country) areas*
District/bioregion/ province/global	- Representative, intact watersheds	- Representative undisturbed watersheds

Bold italics indicate potential conflicts between the two sets of ideals.
Sources: USDA Forest Service, 1974; BCMoF, 1995; Bradley, 1996; Kaplan *et al.*, 1998.

One of the underlying problems with a narrow approach to visual quality is that many people's vision of nature is very different from what nature really can be. In the dominant Western culture, where the general public lives mainly in metropolitan areas, landscape change in the natural environment is usually considered bad (Goodey, 1986). There are often historical justifications for such a view, where the popular culture recognizes previous industrial-scale land exploitation as damaging, as in BC today where forestry and foresters can remain very unpopular.

Therefore, the prevailing scenic aesthetic can be seriously out of step with an ecosystem-management based approach, which promotes man-made activities that emulate natural disturbance, leads to messy foreground debris and clutter, and contemplates large scale harvesting. Attendees at the Peter Wall Institute for Advanced Studies Exploratory Workshop heard very clearly from landscape forestry practitioners that these may not be over-generalizations (Marc, 1999). Thus, while the scenic aesthetic can be criticized for being too narrow, supported primarily by research on the single issue of visual quality, and with very little hard evidence of the ecological benefits of good visual management, it does appear to be in tune with the dominant set of current public perceptions.

Having said this, there are also strong criticisms of the prevailing VRM approach from both sides, particularly in the area of *hiding* forestry behind leave strips (see Chapter 5 this volume). The scenic aesthetic applied in this manner does not sit happily with the realities of backcountry recreation and ecotourism, aerial sight-seeing, and the concept of existence values regardless of whether or not people actually see the landscape in question.

3 The Ecological Aesthetic

More in tune with the new forestry of ecosystem management is the idea that aesthetic appreciation should be informed by ecological knowledge, so that what is good ecologically also looks good to us. This theory draws on Leopold's philosophy of the land ethic (Leopold, 1949) and related arguments from Carlson (1979; 1984), Callicott (1987), and Gobster (1995). Gobster (1999; see also the Foreword to this volume) has recently promoted this theory, which finds support from forest ecologists (*e.g.* Kimmins, 1999 and Chapter 4 in this volume). Presumably informed by more quantitative evidence of ecological condition, and taking a longer-term view of the temporal dynamics of forests, such an aesthetic would avoid the need for managers to preserve the visual intactness of scenic landscapes, or impose, for example, the open ground plane of the savannah hypothesis (Orians, 1986) on the dense rain forest.

3.1 Theoretical Underpinnings of the Ecological Aesthetic and its Relationship to Sustainability

This new aesthetic paradigm seeks explicitly to reconcile aesthetics and ecological sustainability. It advocates a deeper beauty, informed by meaning: it is "as cerebral as it is perceptual" (Flader and Callicott, 1991). Gobster sees the ecological aesthetic as recognizing "the more subtle, experiential, and dynamic qualities that often characterize forest ecosystems of high biological integrity" (Gobster, 1995, p. 7).

The development of such a new aesthetic and its dispersion throughout the population might be analogous to the altered public perceptions of wilderness and wetlands in more recent times, where both have come to be valued in their own right rather than being seen as places to be feared or wasteland to be put to a higher use. The changing nature of some aspects of aesthetic perceptions through history has indeed been amply documented (*e.g.* Porteous, 1996, and Chapter 9 this volume).

Under this theory, knowledge of ecological function and process can overcome what are presumably cultural biases against so-called *messy* ecosystems: those which lack the orderly look of more conventionally managed landscapes. Alternatively, the theory argues that ecological knowledge can broaden peoples vision to recognize the dynamic nature of healthy ecosystems, and to accept landscape change, rather than to adhere to the normal public preference for stability. In recent times, the televized devastation and subsequent natural restoration of Mt. St. Helens in Washington State and the fires in Yellowstone National Park must have contributed significantly to public understanding of ecological processes.

3.2 Problems and Limitations of the Ecological Aesthetic

While this theory appears to have considerable merit for anyone who believes in advancing the cause of ecological sustainability, its practicality is far from being proven. How can such a theory be demonstrated on the ground and disseminated through Western (or other) cultures?

There are a number of concerns regarding this theory. First, we are by all accounts a long way from understanding forest ecosystem dynamics and the spatial principles of landscape ecology: it is therefore hard to define what is actually "good" in a landscape. What if the ecologists are proven wrong in the long term? Is it conceivable that the conventional scenic aesthetic may turn out to be just as ecologically beneficial (at least in a forested landscape setting), after so much energy has been expended on converting people's opinions? Secondly, the theory fails to take into account people's instinctive, genetically-programmed reactions. If much of people's preferences for landscapes is driven by evolutionary biological factors, as suggested by Appleton (1975), then the success of educational (*i.e.* cultural) programmes is likely to be very limited. Lastly, even if the knowledge component can be distributed to the landscape viewers and the cultural component of aesthetic preference proves to be dominant, it is quite possible that people's emotional attachment to place, or other "selfish" motivations, may preclude the spread of the ecological aesthetic. A property owner might say "I know I should burn my forest, but I want to extract the timber instead," or "I want to preserve the existing view for my enjoyment in retirement." Of course, all aesthetic concerns, however they are based, can potentially be overridden by other motivations, but it seems possible that a peculiarly ecological view, which emphasizes broad and long-term future benefits over tangible current gratification, may be more prone to being overridden than other aesthetic concerns.

There are also a number of practical inadequacies in such a theory, around the issue of whether such a new aesthetic could be broadly adopted by the public. Analogies such as wilderness or wetlands may not be appropriate, since neither of these now are normally considered to be *working landscapes* with industrial resource production as a management goal. It is much easier to accept apparentdisorder and disturbance when we are talking about less modified, *naturalistic* landscapes, than when industrial activity is seen as the cause of the disorder and disturbance. Also, the merits of scientific evidence may not be obvious to many communities where academics are seen as out of touch or, worse, in the grip of industry sponsors (Brand, 1999), or where the regulatory and industry bodies are simply not trusted, as is the case in many parts of BC (Klenner, 1999, pers. comm.). The very subtlety and complexity of the ecological relationships may be hard to explain to many people. The time-scales for landscape restoration may simply be too long for the "I want it all now" generation, relative to their typical habitation period in a given location or even their own lifetime. The theory may simply run counter to the rising tide of nostalgia and the increasing value of stability in an urbanizing world with constantly increasing rates of change (Goodey, 1986).

4   A Theory of Visible Stewardship

In response to the shortcomings of the above aesthetic theories, a refined theory linking sustainability and aesthetics is proposed here as a partial explanation for adverse public reactions to forest management practices which ecologists might wish to promote. What we can call a theory of *visible stewardship* adds a key missing ingredient to the ecological aesthetic for working (human-modified) landscapes: *that, other things being equal, we find aesthetic those things that clearly show people's care for and attachment to a particular landscape; in other words, that we like man-modified landscapes that clearly demonstrate respect for nature in a certain place and context.*

This theory emphasizes not whether the landscape looks natural, or orderly, or culturally appropriate, or controlled, so much as whether it looks as though *real individuals care for the land or place*: people who are linked to it, rooted in it, invested in it, working in it in a respectful, symbiotic, and continuously vigilant manner, perhaps even from generation to generation. It can therefore be postulated that forest management activities will not be perceived as good forestry if they fail to demonstrate an obvious and sustained commitment of people to the places under their control (what we might term *visible respect for nature or place*).

4.1 Theoretical Underpinnings of the Visible Stewardship Aesthetic

This theory supplements, rather than competes with the scenic and the ecological theories of aesthetics, as described above, It recognizes the importance of place, as the combination of natural and cultural factors, including spiritual values. It acknowledges people's emotional attachment to, and even their romantic views of, the land. This is consistent with the view expressed and endorsed in another workshop on scenic resources (Strain and Sheppard, 1997), that aesthetic management "puts the heart into resource management." The theory builds upon and supplements the work of Gobster (1995), Thayer (1989), and Nassauer (1995; 1997), among others.

While the theory can incorporate the role of scientific knowledge as described in the ecological aesthetic, it goes beyond the provision of scientific information (such as through interpretive displays and signs) which cites ecological benefits, and embraces certain social concerns. It also goes beyond Nassauer's (1995) recommendations for the design of *cues to care* which attempt to make "messy" ecological practices more orderly, culturally recognizable, and visually acceptable. It requires visible evidence of human attachment to a particular landscape, beyond care as maintenance, but recognizing care as an expression of a deep commitment and affection, with the appearance of fit and hinting of stability. Explicitly tailored to landscapes of human use where at least some human-caused modifications are visible, the theory meshes with Nassauer's description of *vivid care* (Nassauer, 1997), and Thayer's exhortation to make sustainability evident (Thayer, 1989), rather than dressing it up to look like something else.

The idea of visible stewardship as a variable in aesthetic preference can be seen as consistent with several other aesthetic theories: with human habitat theory (Orians, 1986), if we assume that landscape modification practices (such as early agriculture or deliberate burning) had evolutionary survival value; with culturally-based theories of preference, where there is a socially learned association with productive land stewardship or agriculture; or with more recent socio-cultural influences such as a conscious recognition of deliberate progress towards sustainability (*e.g.* as in stewardship awards or certification).

One way to evaluate the merits of this theory is to compare current forestry practices with other large scale human uses of the landscape, such as farming (see for example, Carlson, 1985, and Nassauer, 1989). Both farming and typical forestry in BC, for example, represent large scale, fairly intensive human activities outside urban areas, with an industrial scale of resource extraction, and often leading to dominant rectilinear patterns within a beautiful landscape. Farming in particular appears to be disruptive to both the land surface and any natural vegetation on it. However, landscape transformations in agriculture occur in a readily visible and recognizable annual cycle rather than a forest rotation period closer to a human lifespan. Most significantly, the farmers themselves live there, among their fields, and may have done so for generations (Figure 11.5 and Plate 6). Traditionally, they care for the landscape because they get their livelihood from

Fig. 11.5. The farmhouse as a visible symbol of longstanding stewardship (see Plate 6).

the land, it is their place, and in the popular imagination at least, they love the land. The fact that, in many agricultural landscapes from the Cotswolds to the Corn Belt, farming is becoming more like forestry as corporations consolidate family farms and work the lands with hired labour from elsewhere (Smiley, 1997), may not yet threaten the public's agricultural stewardship aesthetic. We expect to see the farmers' presence (or at least the symbols of their presence, such as farmhouses and harvesting equipment), they are seen to be linked to the land, and we recognize their stewardship.

By contrast, there is little sign that loggers love the place they are logging. Their livelihood comes from the timber, not the place. Foresters do not generally live in their forests as farmers live on their land. They may visit a place once, log it, and leave it, never to return. This is not to say that loggers or forest companies do not actually care about their impact on the land or about the place in which they are working - as Tindall's surveys illustrate (Chapter 5 in this volume), loggers show great pride in doing a good job - but the visible evidence of care for the place itself is largely missing. While in fact there may be a succession of activities in the forest prior to timber harvesting (*e.g.* pre-commercial thinning, timber-cruising, road building, flagging, etc.), these may not be very visible or of long duration. Without a more permanent sign of active management, the harvesting activities in the forest can appear sudden, drastic, and extractive: taking without giving. Especially when geometric clearcuts are left, anyone can *see* that forestry is not respectful of nature (Figure 11.6 and Plate 8), and that the forest management is

Fig. 11.6. Geometric clearcuts seen in middleground symbolize insensitivity to the local landscape, but demonstrate a high degree of coherence and orderliness (see Plate 8).

not sensitive to the local environment. Instead, the observer might assume that the operation was planned in an office many miles away, without concern for what people might think or feel when they see it, with profit as the only motive.

We can extend this comparison between conventional forestry and farming to other land uses, as in Table 11.2. Here, we address a range of land uses, from the urban shopping mall to the designated wilderness area, and postulate an overall expected aesthetic preference or perceived image (in Western cultures) for each, based on the author's practical experience and very broad generalizations from research on perceptions of various land uses (*e.g.* Zube *et al.*, 1982; Kaplan *et al.*, 1998). For example, shopping malls are usually not seen as aesthetically beneficial, whereas farmland and protected wilderness are. We can also relate these land uses to various individual aesthetic criteria used in landscape research, such as perceived naturalness, coherence, and orderliness, as well as the proposed criterion of visible stewardship. It is argued here that the overall image (good or bad) of these land uses correlates most closely to the visible stewardship criterion: gardens, farms, and designated wilderness tend to have the most positive image, and these are the land uses associated most strongly with obvious evidence of care and protection (in the case of wilderness, care is evinced through special levels of protection, management by parks staff, and the use of signs). In these landscapes at least, the following assertion may be correct:

> The aesthetic of care incorporates change quite easily, if change appears to be mediated by human intention. A well-cared for landscape is expected

Table 11.2. Land uses evaluated on selected aesthetic criteria and overall preference.

Aesthetic criteria	Land uses (Visible human activity on the land)						
	Residential gardens	Shopping mall	Farm land	Conventional forestry	"New forestry"	Community forestry	Designated wilderness
Perceived naturalness	No	No	No	Yes/no	Yes/no	Yes/no	Yes
Coherence	Yes	Yes	Yes	Yes/no	Yes/no	Yes/no	Yes/no
Orderliness	Yes	Yes	Yes	No: foreground Yes: middleground	No: foreground Yes/no: middleground	Yes/no: foreground Yes/no: middleground	Yes
Control of nature	Yes	Yes	Yes	Yes	Yes/no	Yes/no	No
Visible stewardship	Yes	No	Yes	No	No	Yes	Yes
Overall image (preference)	Good	Bad	Good	Bad	Bad?	Good?	Good

Forestry definitions:

Conventional Forestry: forestry with clearcutting in regular-shaped blocks of relatively large size, *e.g.* 200 acres, and traditional commercial forestry practices as practised in the 1970s and 1980s in BC, with the emphasis on yield maximization.

"New Forestry": forestry with patch cuts, variable block size, and retention of snags and large woody debris, as required under the BC Forest Practices Code, with the emphasis on ecosystem management.

Community Forestry: forestry in the vicinity of a permanent community, managed locally by local residents and foresters to balance commercial, ecological, and community needs.

to change, but it is expected to exhibit signs that well-intentioned people are watching over that change (Nassauer, 1997, p. 74).

We can consider forestry's place in this scheme. Despite being coherent and orderly, at least in non-foreground views (Figure 11.6), conventional forestry practices are generally not favoured (*e.g.* Bradley, 1996; BCMoF, 1996; and Paquet and Belanger, 1997). However, so-called *new forestry* which seeks to emulate natural processes through ecosystem management, may not fare any better in the public eye, as discussed above. Neither of these forms of management typically exemplify visible stewardship, beyond the use of signs that typically document which company is managing the forest and when the replanting took place. In North America, at least, the jury is still out on whether real community forestry, which might be expected to demonstrate more obvious stewardship, will escape the negative associations commonly held by the public with regard to other forms of forestry.

4.2 Relationship to Sustainability

The theory of visible stewardship argues that if the visible evidence of stewardship is based on real stewardship (and not just the superficial appearance of good management), then we get the best of both worlds: a sustainable forest and a preferred landscape. If sustainability is defined broadly as having some sort of long term stability in human terms (not necessarily a steady-state ecological environment), one would expect a strong association between landscapes which sustain communities and those which are preferred. This is especially true where the landscape itself becomes the visible evidence of the success and benefits of stewardship over the long term. Put another way, the theory should deliver a managed environment which functions as described in the ecological aesthetic, but which yields human responses more like those of the scenic aesthetic.

4.3 Problems and Limitations of the Visible Stewardship Aesthetic

In justifying the theory by comparing forestry with other land uses such as farming, a possible threat to validity is that farmed landscapes seem to be within the community, or at least, within a settled landscape; whereas forestry is often thought of as *outside* the community, amidst raw nature, as being *out there*. Many writers have expressed the historically pervasive view of the forest as mysterious, wild, and threatening: for example, H.E. Bates (1936) contrasts the warmth and bustling human activity of the mid-summer farmland of England with the darkness and remoteness of an adjoining thick wood:

> In the wood, on the fiercest of noons, there is a coldness and stillness and shadowiness under the trees that is like a momentary oasis of death. For a moment or two, after the sickening blaze of sunlight, the wood is

a relief, but before long that utter completeness of shade begins to seem unfriendly, then forbidding, and finally hostile. It is too complete. It is like the fore-thrown shadow of winter. (p. 84).

In more recent times, when public sentiment has swung behind the idea of natural landscapes as relic islands in a sea of people, rather than the matrix around communities, forests are still typically perceived as being separate. Therefore one might expect human activity in the forest to be seen as automatically more disruptive of nature than in the more obviously used and manipulated farming landscape: forestry as industry, out of place in the wilderness.

There are, however, an increasing number of people in places who choose to live *in* the forest, be it the log-and-stone palaces of Whistler and Lake Tahoe, or the owner-built cabins of the Slocan Valley in BC. This can be taken as further evidence of the cultural trend away from forests as hostile no-go areas; perhaps then, in at least some urban influence areas, the forests may be seen by an increasing number of people as harbouring or even forming communities, rather than existing outside them. Nevertheless, this is still some way from a full acceptance of the working (*i.e.* productive) forest as being within the community, rather than simply a recreational, *passive* forest within the community. Indeed, in places like the Slocan Valley and Whistler, the *newcomers* to forest communities may actually intensify the conflict over local harvesting activities being pursued by the traditional residents.

There is of course a wealth of research which indicates that people prefer natural landscapes over man-modified landscapes (see Zube *et al.*, 1982; Kaplan *et al.*, 1998), which would argue against a theory of visible stewardship. Anderson (1981) has shown that people even prefer forest landscapes labelled as natural over identical scenes labelled as managed forest. However, as previously discussed, it is also well documented that people like certain kinds of productive, man-modified landscapes (*e.g.* Nassauer, 1988; USDA Forest Service, 1995).

It can also be argued that the real reasons for people's dislike of forestry in the landscape are largely social, rather than derived from signs in the landscape itself. Causes of dislike of forestry, which may have little to do with the local level of visible stewardship, include the global image of *bad* versus *good* forestry, as communicated in much environmental literature (see for example, the Sierra Club's coffee-table book entitled *Clearcut* (Devall, 1993). There may also be strong negative reaction to the lack of involvement of the affected parties in the decision-making process for forestry, as suggested by Kruger in Chapter 12 of this volume. These considerations would apply to any directly aesthetic theory of preference, not just the visible stewardship theory. However, at the local level, visible stewardship may offer a stronger tie to such social processes than the competing theories: providing a more visible presence on the land may naturally lead to more interaction with (and influence from) the local community and visitors.

Clearly, one of the limitations is the lack of hard evidence for such a theory, coupled with the scarcity of precedents in forestry practice that exist in North America. There has been considerable landscape perception research focused on *immediate* reactions to landscape conditions; there has been comparatively little research in the USA and Canada on the cumulative, longitudinal responses to management. Such research would be the most relevant to a sustained programme of visible stewardship.

From a practical standpoint, the delivery of visible stewardship in a large scale forest setting may be seen as simply not cost-effective under current economic and tenure arrangements. Hiring additional forest stewards to conduct regular activities in and around forest communities may not be economically justifiable, unless intensive forest management is to be practised (still a rarity in places such as British Columbia). There is also a risk that such people might be viewed as a token presence, an advertisement for sustainable management but without real power over local decisions. It is possible, however, that, much as with community policing, the initial outlay may more than pay for itself through reduced costs for obtaining approvals, avoiding lawsuits, etc. The brief history of community forestry in BC has already demonstrated considerable advantages in this area (Allen and Frank,1994).

5 Implications for Practice and Research

5.1 Implications for Practice

Conventional VRM seeks to hide the work of the forester: forest landscape designers are therefore most successful when their work cannot even be detected. Are there not ways to celebrate such careful attention and design, or provide a more visionary and locally distinctive approach? The theory of visible stewardship suggests that acceptable forestry needs to provide *more evidence of human activity*, not less. Intensity forestry, for example, should be acceptable and even desirable, if appropriately located, managed and designed. Under this hypothesis, standard VRM techniques may serve to identify thresholds where forestry practices become apparent to observers, and may help to achieve more sensitive designs at some scales, but would otherwise make limited contributions to the public image of a well-loved, well-stewarded forest.

Instead, design should develop clear *cues to care* (Nassauer, 1995) which are appropriate to wilder forested landscapes; sensitive to human preferences for natural, orderly, and biologically preferred landscape conditions; but also demonstrating an obvious attachment to the land. This can be seen as consistent with community forestry, forestry on private land, and even long-term area-based forest tenures. However, more than these measures is needed: there should be a visible and continuous means of public involvement on-site via known and approachable individuals, coupled with an open social process for decision-making and steady feedback of monitoring results to demonstrate the fruits of stewardship.

Programmes for directly involving local people in the management and monitoring of forests on the ground, as described by Nassauer (1997) and exemplified in community-based forest restoration projects such as the Wychwood project in the UK (Corbett, 1998), may lead to greater understanding and acceptance of careful management.

Forest companies need to be more visible, have a stake in the land and the local community, with commitment of real people devoted to managing individual watersheds or landscape units. It is particularly important that such programmes represent much more than public relations exercises, and that they do more than placing a human face on otherwise unchanged priorities and practices. Ecologists and landscape architects both need to be more prominent in the community, and fully prepared to engage in discussions with the public on all issues, including those on the *subjective* and emotional side, such as aesthetic and spiritual values.

New approaches in Canada, such as locally-based community forestry pilots in BC and discussions on First Nations' forest tenures, appear to offer hope for a more successful union of aesthetic and spiritual values (broadly defined) with ecological integrity, through visible stewardship. In both these situations, the direct tie between what happens on the land and the fortunes of the local community (short-term and long-term), provides potentially fertile ground for enhanced public understanding of ecological processes and temporal consequences. Also, the visible, permanent, and local presence of the forest manager(s) and decision-makers within the community makes possible the development of trust, perceived inter-dependency between the stakeholders, and local pride in the careful use of the resource (Allen and Frank, 1994).

5.2 Implications of the Visible Stewardship Theory for Research

There is a surprising lack of studies that explore this and other plausible theories linking sustainability to aesthetics, even though considerable data exists which appears to be suitable for research comparing ecological conditions (*e.g.* downed timber) to visual values. Caution is urged, however, regarding the over-use of static photographs which are snapshots in time, to represent a complex landscape with considerable spatial and temporal variation and viewing dynamics. More experiential settings and stimuli, including real-world landscape walks and perhaps immersive, dynamic visualizations (as described by Sheppard, 2000 and by Danahy in Chapter 15 of this volume), should be used in response testing, to bring subjects closer to real-life decision-making (Gobster, 1999). We should test the effects of not only forest practices, but also of place, the types of people involved, and the process of decision-making. Research should explore the differences between rooted communities (living in and impacted by local landscapes) and the more urban, heterogeneous populations often used in perception testing.

In particular, research is needed in: quantifying observed relationships between aesthetic character/preference and various ecological and management indicators of sustainability, going beyond those shown in Table 11.1; more comprehensive assessment of public and stakeholder attitudes to test alternative theories and develop a more robust basis for balancing ecological and aesthetic goals; and development of aesthetic and place-based indicators of forest sustainability which are more defensible and recognizable.

6 Conclusion

As the principles of ecosystem management have come to dominate forestry and industry shifts towards more sustainable (*i.e.* certifiable) forestry, broader questions have emerged: does visual resource management support or conflict with ecosystem management goals, and vice versa? If there is conflict between these goals, how can they be united?

There appear to be real conflicts between public perceptions (based on visual indicators) and ecological sustainability in some circumstances, with potentially serious consequences for timber supply, biodiversity, tourism, and related economic effects. Visual resource management as it has been practised over the last quarter century has in general successfully preserved the most sensitive front-country landscapes, but now appears unable to solve the conflicts raised by the prospect of ecologically sustainable, industrial scale forestry in these more visible locations.

The gap between sustainable forestry and public perception or social acceptance cannot be bridged by the narrowly defined scenic aesthetic (appraised largely by experts), nor by the ecological aesthetic as conveyed by the limited flow of expert-derived ecological information, typically delivered in signs, extension programmes, and public relations programmes. A broader approach is needed, integrating aspects of both these aesthetic creeds, but more inclusive of social goals as well as deep-rooted reactions to landscapes and cultural expectations.

The theory of *visible stewardship* is advanced to meet this supplemental need; it postulates that forest management activities will not be perceived as good forestry if they fail to demonstrate an obvious and sustained commitment of people to the places under their control (*i.e.* visible respect for nature and place). Without clear signs of people's attachment to, affection for, and protection of the landscape, the best management intentions may be doomed to fail if they are visible to the public: in short, if they send the *wrong message*. A more acceptable and appropriate message would convey that the people working on the forest live locally, amid a changing but carefully managed, productive, and ecologically sound forest. The accent should be on a clear *fit* with nature and culture, showing continuity of management rather than discontinuity. The message should also convey the presence of an open social process for management, through appropriate social indicators of sustainability, preferably personified in the shape of a forest steward living in the community and monitoring the health of the forest environment.

While none of this argues against the need for environmental education to expand public understanding of forest management issues in relation to ecological sustainability, it is difficult to believe that we can or should focus most of our energies on changing people's minds primarily through conventional means such as extension education or public relations. Rather, the principles of visible stewardship suggested here may provide the most effective forum for developing a broader, ecological aesthetic. At the same time, visible stewardship may provide an opportunity for the role of VRM-based design to be freed from its shackles and expanded. Embracing this combination of theoretical approaches to the perception of sustainable landscapes may ultimately represent the greatest strength of visible stewardship movement.

References

Allen, K., and D. Frank. (1994) Community forests in British Columbia: models that work. *Forestry Chronicle* 70(6): 721-724.

Anderson, L. (1981) Land use designations affect perception of scenic beauty in forest landscape. *Forest Science* 27: 392-400.

Appleton, J. (1975) *The Experience of Landscape*. John Wiley and Sons, Chichester.

Bates, H.E. (1936) *Through the Woods: The English Woodland - April to April*. Victor Gollanz Ltd., London.

Bradley, G.A. (1996) *Forest Aesthetics-Harvest Practices in Visually Sensitive Areas*. Washington Forest Protection Association, Olympia, WA.

Brand, D. (1999) *Sustainable Ecosystems, Public Attitudes and the Formulation of Public Policy*. Paper presented at the Peter Wall Institute for Advanced Studies Exploratory Workshop, Linking Sustainability and Aesthetics: Do people prefer sustainable landscapes? February 24-28, 1999, UBC, Vancouver.

Callicott, J.B. (1987) The land aesthetic. In: J.B. Callicott (ed) *Companion to a Sand County Almanac: Interpretive and Critical Essays*. University of Wisconsin Press, Madison.

Carlson, A. (1977) On the possibility of quantifying scenic beauty. *Landscape Planning* 4: 131-172.

Carlson, A. (1979) Appreciation and the natural environment. *Journal of Aesthetics and Art Criticism* 37: 267-276.

Carlson, A. (1984) Nature and positive aesthetics. *Environmental Ethics* 6: 5-34.

Carlson, A. (1985) On appreciating agricultural landscapes. *Journal of Aesthetics and Art Criticism* 24: 301-312.

Carlson, A. (1998) Aesthetic appreciation of nature. In: E. Craig (ed) *Routledge Encyclopedia of Philosophy, Vol. 6*. Routledge, London.

Corbett, J. (1998) Creating a new conservation vision: the stakeholders of Wychwood. *ECOS*, 19(2): 20-30.

Crowe, S. (1966) *Forestry in the Landscape Forestry Commission Booklet No. 18*. Her Majesty's Stationery Office, London.

Daniel, T.C. and R.S. Boster. (1976) *Measuring Landscape Esthetics: The Scenic*

Beauty Estimation Method. USDA Forest Service Research Paper RM-167. USDA Forest Service, Rocky Mountain Forest and Range Experiment Station, Fort Collins, CO.

Devall, B. (ed) (1993) *Clearcut: the Tragedy of Industrial Forestry*. Sierra Club Books/Earth Island Press, San Francisco.

Diaz, N. and D. Apostol. (1992) *Forest Landscape Analysis And Design: A Process for Developing and Implementing Land Management Objectives for Landscape Patterns*. Report No. R6 ECO-TP-043-92. USDA Forest Service, Pacific Northwest Region, Portland, OR.

Flader, S.L. and J.B. Callicott (eds) (1991) *The River of the Mother of God, and other essays by Aldo Leopold*. University of Wisconsin Press, Madison.

Gobster, P.H. (1995) Aldo Leopold's ecological esthetics: integrating esthetic and biodiversity values. *Journal of Forestry*, February 1995: 6-10.

Gobster, P.H. (1999) An ecological aesthetic for forest landscape management. *Landscape Journal* 18(1): 54-64.

Goodey, B. (1986) Spotting, squatting, sitting, or setting: some public images of landscapes. Chapter 6. In: E.C. Penning-Rowsell and D. Lowenthal (eds) *Landscape Meanings and Values*. Allen and Unwin, London.

Kaplan, R., S. Kaplan and R.L. Ryan (1998) *With People in Mind*. Island Press, Washington, DC.

Kimmins, J.P. (1999) Biodiversity, beauty, and the "beast": Are beautiful forests sustainable, are sustainable forests beautiful, and is "small" always ecologically desirable? *Journal of Forestry*, 75(6): 955-960.

Leopold, A. (1949) *A Sand County Almanac*. Oxford University Press, Oxford.

Litton, R.B. Jr. (1968) *Forest Landscape Description and Inventories*. USDA Forest Service Research Paper PSW-49. USDA Forest Service, PSW Forest and Range Experiment Station, Berkeley, CA.

Marc, J. (1999) *Ecological and Aesthetic Relationships*. Paper presented at the Peter Wall Institute for Advanced Studies Exploratory Workshop, Linking Sustainability and Aesthetics: Do people prefer sustainable landscapes? February 24-28th, 1999, UBC, Vancouver.

Nassauer, J.I. (1988) The aesthetics of horticulture: neatness as a form of care. *Horticultural Science* 23: 937-977.

Nassauer, J.I. (1989) Agricultural policy and aesthetic objectives. *Journal of Soil and Water Conservation* 44: 384-387.

Nassauer, J.I. (1995) Messy ecosystems, orderly frames. *Landscape Journal*, 14(2): 161-171.

Nassauer, J.I. (1997) Cultural sustainability: aligning aesthetics and ecology. In: J.I. Nassauer (ed) *Placing Nature: Culture and Landscape Ecology,* Chapter 4, 65-83. Island Press, Washington, DC.

National Environmental Policy Act of 1969 (1969) Debate on Conference Report Number 765, Congress Records, 91st Congress, 1st Session, 1969, 115, 40415-27.

Newby, F. (no date) *Environmental Impact Appraisal of Proposed Developments in the Harney Peak Area of the Black Hills.* USDA Forest Service Pacific Southwest Forest and Range Experiment Station, Berkeley, CA.

Orians, G.H. (1986) An ecological and evolutionary approach to landscape aethetics. Chapter 2 In: E.C. Penning-Rowsell and D. Lowenthal (eds) *Landscape Meanings and Values.* Allen and Unwin, London.

Paquet, J. and L. Belanger (1997) Public acceptability thresholds of clearcutting to maintain visual quality of boreal balsam fir landscapes. *Forest Science* 43(1): 46-55.

Picard, P. and S.R.J. Sheppard (2000) Timber supply and aesthetics at the landscape level: Theoretical relationships and potential win-win solutions. *Journal of Ecosystems and Management* (Southern Interior Forest Extension and Research Partnership). In press.

Porteous, J.D. (1996) *Environmental Aesthetics.* Routledge, London.

Province of British Columbia, Ministry of Forestry (BCMoF) (1981) *Forest Landscape Handbook.* Information Services Branch, BCMoF, Victoria BC.

Province of British Columbia, Ministry of Forests (BCMoF) (1995) *Visual Landscape Design Training Manual.* Recreation Branch Publication 1994:2. Recreation Branch, BCMoF, Victoria BC.

Province of British Columbia, Ministry of Forests (BCMoF) (1996) *Clearcutting and Visual Quality: A Public Perception Study.* Recreation Section, Range, Recreation and Forest Practices Branch, BCMoF, Victoria BC.

Province of British Columbia, Ministry of Forests (BCMoF) (1997a) *Visual Landscape Inventory Procedures and Standards Manual.* Forest Practices Branch, BCMoF, Victoria BC.

Province of British Columbia, Ministry of Forests (BCMoF) (1997b) *Managing Visual Resources to Mitigate Timber Supply Impacts.* Draft Memorandum from Henry Benskin, Director Forest Practices Branch, to Larry Pederson, Chief Forester. Forest Practices Branch, BCMoF, Victoria BC.

Province of British Columbia, Ministry of Forests (BCMoF) (1998) *Procedures for Factoring Visual Resources Into Timber Supply Analyses.* Forest Development Section, Forest Practices Branch, BCMoF, Victoria BC.

Rountree, R. (1998) Modeling fire and nutrient flux in the Lake Tahoe basin. *Journal of Forestry,* 96(4): 6-11.

Sheppard, S.R.J. (1999) *The Visual Characteristics of Forested Landscapes: a Literature Review and Synthesis of Current Information on the Visual Effects of Managed and Natural Disturbances.* Technical Report prepared for the BC Ministry of Forests, Kamloops, TELSA modeling Project, Adams Lake Innovative Forest Practices Agreement (IFPA).

Sheppard, S.R.J. (2000) Visualization as a decision-support tool in managing forest ecosystems. *Compiler* 16(1): 25-40.

Smardon, R.C. (1986) Review of agency methodology for visual project analysis. Chapter 9. In: R.C Smardon, J.F Palmer, and J.P. Felleman (eds) *Foundations for Visual Project Analysis*, John Wiley and Sons, New York.

Smiley, J. (1997) Farming and the landscape, Chapter 2. In: J.I. Nassauer (ed)

Placing Nature: Culture and Landscape Ecology. Island Press, Washington, DC, pp. 33-43.

Strain, R.A. and S.R.J. Sheppard (eds) (1997) *Proceedings of Establishing the Worth of Scenic Values: the Tahoe Workshop, October, 1997, South Lake Tahoe, NV.* Stateline, NV: Tahoe Regional Planning Agency. <http://www.ceres.ca.gov/trpa/SCENIC.pdf>.

Thayer, R.L. Jr. (1989) The experience of sustainable landscapes. *Landscape Journal* 8(2): 101-110.

United States Department of Agriculture Forest Service (1974) *National Forest Landscape Management, Vol.2, Chapter 1: The Visual Management System.* USDA Agriculture Handbook No. 462. US Government Printing Office, Washington, DC.

United States Department of Agriculture Forest Service (1981) *Use of "Effective Alteration" in FORPLAN — Visual Resource in the Forest Plan.* Document sent to the Forest Supervisors from Zane G. Smith, Jr., Regional Forester on August 28, 1981. Forest Service, United States Department of Agriculture.

United States Department of Agriculture Forest Service (1995) *Landscape Aesthetics: A Handbook for Scenery Management.* USDA Forest Service Agriculture Handbook No. 701. US Government Printing Office, Washington, DC.

Zube, E.H., J.L. Sell, and J.G. Taylor (1982) Landscape perception: research, application and theory. *Landscape Planning* 9: 1-33.

Chapter twelve:

What is Essential is Invisible to the Eye: Understanding the Role of Place and Social Learning in Achieving Sustainable Landscapes

Linda Kruger
USDA Forest Service, Pacific Northwest Research Station
Seattle, Washington

1   Introduction

In the search for management practices that achieve sustainable landscapes there is more to gaining public support or acceptance than achieving some visual aesthetics. Questions that provide a focus for this chapter are "How do people experience and understand landscapes and how does this relate to sustainability?" and "What processes are available to identify what to sustain and how?"

Various disciplines have taken differing approaches to these questions. Landscape architects and environmental psychologists have focused on measuring scenic beauty and aesthetic preference (Daniel and Boster, 1976; Nassauer, 1983; Gobster and Chenoweth, 1989). Social ecologists, forest, natural resource, and environmental sociologists, cultural geographers, and environmental historians have explored what it is that determines whether forestry practices and conditions are socially acceptable (Shannon, 1991; Brunson, 1996; Stankey, 1996; Shindler and Cramer, 1999). Rather than focusing only on the physical attributes of landscape, this paper approaches landscapes as places with meaning and significance that people experience holistically. If our only relation to a landscape was based on our visual perception, then aesthetic value might be adequate in helping us understand whether people would prefer more sustainable landscapes over less sustainable ones. By thinking of landscapes as places, our expectations of how a place is managed, and our judgements regarding the acceptability of that management, become more complex. This experiential aspect has important implications for how we study, plan for, and manage landscapes for sustainability.

Sustainable forestry as a social and political concept requires public dialogue to define what should be sustained and what tradeoffs will be made to achieve sustainability. With sustainability as a goal of natural resource management, it is imperative to involve citizens in decision making processes. However, how resource managers feel about the role of the public in planning and decision making processes can work to either limit or expand opportunities for public involvement.

The first section of the paper provides a brief discussion of the concept of sustainability. This discussion is followed by an introduction to landscape as place. Next the concept of public philosophy and its implications for sustainability is presented. The final section of the paper focuses on social learning theory and involving citizens in planning and decision making as social learning processes. These processes provide opportunities to expand meaningful deliberation among scientists, managers, and citizens through collaborative learning activities.

2 Sustainability

The term sustainability is being used with increasing frequency in discussions of management and use of natural resources. Task forces and groups have formed worldwide, at scales ranging from the local community to multinational groups, to develop sustainability criteria and indicators appropriate at various scales (*e.g.* Sustainable Seattle and Pierce County efforts in Washington and the Montreal Protocol).

Although there is fairly universal agreement on the desirability of sustainability, there is much ambiguity and little consensus exists about exactly what the concept means and there are many definitions. Specifically, what should be sustained? Who gets to decide what to sustain? How is sustainability achieved? How is progress or success measured? Who will win and who will lose? The most common definition of sustainability derives from the Brundtland report (World Commission on Environment and Development, 1987). Based on this report, sustainability has come to mean meeting the needs of today without jeopardizing the ability of future generations to meet their needs.

This definition is itself ambiguous, and the ambiguity is in part why sustainability can be so appealing to such a variety of audiences. It is what Shumway (1991) refers to as a "guiding fiction", which means that although sustainability provides a focus that people can rally around, accomplishing sustainability cannot be measured or proven. Thus, it serves as a catalyst, creating a sense of community and purpose around which people can organize for discussion and action (McCool and Stankey, 1998). Deciding what to sustain and how to implement sustainable practices are locally particular and socially constructed, meaning they are defined by decision makers.

Sustainability of timber yield has far different implications than sustainability of a whole forest ecosystem, or simultaneous sustainability of forests and human communities. Yet each is referred to in terms of sustainability. The concept is used in three ways: in relation to a single resource such as timber; in the larger, more inclusive context of an ecosystem; and finally to embrace social, cultural, and economic aspects of ecosystems (Dixon and Fallon, 1989; Gale and Cordray, 1994; Sancar, 1994). This third concept recognizes that a biophysical perspective is too simplistic and does not account for influences such as shifts in societal values for, and uses of, resources, policy changes, lawsuits, and other social, cultural, and economic influences. This broader perspective seems to be the most widely accepted.

Thus, by using the most widely accepted definition, derived from the Brundtland Report, sustainable processes value humans and non-humans equally with a concern for equity, justice, and human rights, along with environmental well-being. If a practice is not socially or economically sustainable, it cannot be sustainable in an ecological sense (McCool and Stankey, 1998). As mentioned earlier, sustainability itself cannot be measured; however we can measure the progress made in applying sustainable practices. Attempting to measure the success of sustainable practices requires paying attention to conditions, traditions, perceptions, and values of local citizens and communities as well as characteristics and conditions of the biophysical system.

Sustainability therefore is a fundamentally moral and value-oriented issue (McCool and Stankey, 1998) and cannot successfully be addressed as if it were simply a scientific or technical issue. This has two implications. First, it means that although science can develop ways to help achieve objectives that are identified through public processes, help explore the consequences of implementing those choices, and help us better understand relations and dynamics, it is not the role of science to define the objectives. Second, while remaining true to the underlying concept that sustainability means meeting today's needs without jeopardizing the resources needed to meet tomorrow's needs, diverse stakeholders must be involved in determining what to sustain and how (the objectives) and what measures fairly represent success in implementing sustainable practices (McCool and Stankey, 1998). Because of the complex nature of sustainability, forums for social deliberation and shared learning processes are needed to bring people together to explore what to sustain, how to proceed, and potential outcomes and implications of a particular course of action or inaction (Yankelovich, 1991).

People describe, define, and therefore value ecosystems in different ways. We each see, experience, and interpret things differently. Thus, what is sustainable from one perspective may be unsustainable from other perspectives (Orians, 1990). A forest, for example, can be described and experienced by people in many ways. When looking at a forest, a developer might see residential lots and roads, a forester might see the forest in terms of how many board feet could be removed, and local community members might see the forest as a source of community water or recreation or tourism opportunities. Some people experience the forest as a spiritual place, almost a natural cathedral, whereas hunters or fishers may value the forest as habitat for the fish and game species they pursue. Each perspective is equally valid and each can be evaluated in terms of sustainability. How people "see" the forest depends on their identities and how they define themselves in relation to the forest (Greider and Garkovich, 1994). In each case, the forest is the same physical environment, but it has different meanings for different people and even different meanings for the same person at different times.

Landscapes are the symbolic environments created by human acts of conferring meaning to nature and the environment, of giving the environment definition and form from a particular angle of vision and through a special filter of values and beliefs. Every landscape is a symbolic environment. These landscapes reflect our self-definitions that are grounded in culture (Greider and Garkovich, 1994: 1).

Although we can discuss sustainability in theoretical terms, it is only in relation to a specific place that we can discuss the actual implementation of strategies or implications of practices to achieve sustainability. Identification with a specific place provides an orientation around which people with differing perspectives can come together to collectively define issues and problems and build solutions.

At the heart of this collaboration is the recognition that often, although not always, for members to meet their own goals, the cooperation of others is necessary. "Sustainability as a concept... makes explicit otherwise hidden interdependencies" (Social Science Research Group, 1994: 3). When adversaries realize that they cannot achieve their goals without the help of other groups, they often become more willing to find ways to work together. The Applegate Partnership in Ashland, Oregon, is one example of how disparate groups have come together to explore opportunities for sustainability when they recognize they share concerns about a particular place (McCool and Stankey, 1998).

When based on a contemporary definition of sustainability that brings together cultural, social, biophysical aspects, human judgements and values, social action, and holistic management of ecosystems, the implementation of sustainable practices requires not only scientific and technical knowledge but "local, cultural, subjective/experiential, and situational" knowledge (Sancar, 1994: 323). The most common approach in the United States, however, continues to be an expert-driven, empirical, quantitative analysis and large-scale computer modelling effort (Sancar, 1994).

One reason the pursuit of sustainability requires local knowledge and involvement is because people often do not support what they do not understand and do not understand what they have not been involved in (FEMAT, 1993). Defining what to sustain and how to implement practices that help achieve sustainability are social and ethical questions that require that people become involved in decision making processes and the implementation of actions based on their decisions. Because characteristics of specific situations and local knowledge and involvement will differ from case to case, opportunities to apply a template approach (one-size-fits-all) are limited.

A second reason for local involvement is that the implementation of sustainable practices will be unique to a particular place (Sancar, 1994). Sancar (1994: 327) suggests that "the scientific basis for sustainability rests on a critical understanding of evolutionary, historical, cultural, and material aspects of places achieved via long-term experience of locals" (Sancar, 1994: 327). Gathering and documenting local knowledge of landscapes and encouraging and strengthening the local capacity to record and use local expertise is an important aspect of sustainable practices (Sancar, 1994).

3   Forest Landscapes as Places

Landscapes are experienced by people as places - as a combination of setting, landscape, relations with other people and biophysical components, personal experiences, memories, rituals, and in relation to other places (Fishwick and Vining, 1992), and to themselves. A place is more than a physical piece of real estate. Space, the piece of real estate, becomes place through the meanings it is given by people who interact with it over a period. Politicians, professional planners, and managers who make land use and land designation decisions also create places by giving them meaning.

Symbolic meanings define the way individuals and groups relate to a particular place. Embedded in the meanings people attach to places are norms for accepted (appropriate) and unaccepted (inappropriate) behavior. To understand how acceptable a particular activity might be in a particular place, the meanings people have ascribed to the place need to be understood. This factor has important implications for how we study, plan for, and manage forest landscapes. Scientists and managers also overlay biophysical features with meanings. These meanings, although no better or worse, are often different than the meanings ascribed by others who interact with places.

Science traditionally sorts out the feelings, symbolic meanings, and sentiments from the tangible features of our environment. By uncoupling the meanings from the tangible features, the unity and coherence of the experience as a whole is lost (Walter, 1988). This uncoupling has resulted in the division of landscape study into two orientations - the scientific orientation deals with the tangible, physical features (visual indicators), and the arts and humanities orientation deals with feelings, meanings, and sentiments. This separation leads to the scientific assumption that we can get an accurate or adequate (and unbiased or objective) measure of the significance of a landscape by simply evaluating what we can see with our eyes - the visual aesthetic without considering meanings and sentiments.

Rather than depending on what we can see, we might take seriously the admonishment of the fox to the little prince in Saint-Exupery's 1943 children's classic, *The Little Prince*. "It is only with the heart that one can see rightly; what is essential is invisible to the eye." In terms of sustainable forestry, much of what is essential is invisible to the eye. We are asking for trouble if we presume to capture social judgement solely through some measure of a visual aesthetic. What is essential may have more to do with both social and ecological relations and processes than visual outcomes.

Can we assess sustainability through visual analysis? That may depend partly on how we define sustainability and what we hope to sustain. While spatial-visual qualities and landscape forms influence meaning and hold significance for people there are other aspects of places that also influence meaning and significance. If we could define places and the value they have for people by their visual qualities alone, a visual analysis might suffice. Places, however, are much more than visual backdrops or pretty places to recreate. Landscapes are places with meaning and significance based on more than visual qualities. Because of this, there is a pressing need to expand the way we approach the management of natural resources and amenities to explicitly include places and their meanings.

Place provides a focal point for human experience. In addition to physical attributes (aesthetics), place encompasses activities, experiences, memories, individual and group identity, meanings, relations, and what is often referred to as a sense of place or *Genius loci*. Sagoff (1992: 69) implores us to consider landscape as "not just a collection of resources or materials for our use but as the habitat in which we live." He suggests that we cherish places not just for what we can get from them but for the way we define ourselves in relation to them.

Places are significant not solely because of any inherent qualities or discernable facts but because of the meanings and values that evolve through human interaction with and in places (Kruger, 1996). "For each inhabitant, a place has a unique reality, one in which meaning is shared with other people and places. The links in these chains of experienced places are forged of culture and history" (Rodman, 1992: 643).

A landscape is a central part of a local community, often becoming an essential part of a community's identity. "The relationship between community and place is indeed a very powerful one in which each reinforces the identity of the other, and in which the landscape is very much an expression of community held beliefs and values and of interpersonal involvements" (Relph, 1976: 34). People who have recently moved into an area often come to care about the places they live, work and play in as much as people who have lived in a place for a long time. These places become more than forest landscapes. They become places with stories, memories, meanings, sentiments, and personal significance. Tourists and vacationers who return to the same location each year often develop these same relations with a frequently visited place. For this reason, discussions of sustainability need to consider questions such as "How does the community conceive of the landscape?" and "What would maximize landscape value to the community?"

For example, to help researchers understand how farmers define places, and how these definitions influence farm management, researchers in Western Australia have used photos to access meanings farmers ascribe to features, objects, and events (Moore, 1997). As Moore (1997) points out, a strong sense of place does not necessarily lead to sustainable practices; it is likely, however, that a lack of sense of place, or placelessness, is linked to unsustainable practices. Thus, attachment to, or identification with, a place seems to be a necessary but insufficient condition for sustainability (Moore, 1997). Further research is needed to explore the relation between sustainable practices and attachment to or identification with place.

As a society, we need to rediscover the landscape as a cultural system - a place of relations between people and the natural world - by becoming more actively involved in learning about and caring for the landscape. It is this getting to know a place, this developing a sense of care and connectedness (Orr, 1992), that some suggest is necessary for achieving sustainability (Sancar, 1994). Sustainability of both ecological and social systems depends on understanding and respecting the relations between the two systems. This paper contends that this understanding cannot be circumscribed by aesthetic value alone. In addition, the argument is made that the public has an important role in defining what to sustain and how trade-offs will be reconciled. Understanding two dramatically different perspectives of the public's role in decision making may shed light on different approaches to public participation and the underlying philosophy that is needed to facilitate social learning and collaboration.

4 Implications of Public Philosophy

Sandel (1996) contrasts two public philosophies that he believes underlie how the role of the public is conceptualized. The first public philosophy stems from a branch of political theory based on the idea "that we are separate, individual persons, each with our own aims, interests, and conception of the good" (Sandel, 1984: 16). Individual values are of primary importance; therefore a sense of a common good is assumed to take care of itself (Unger, 1984), and the common good is not considered in decision making. The term competitive pluralism conveys this notion of each individual competing with others to satisfy his or her individual desires.

An alternative public philosophy builds on "deliberating with fellow citizens about the common good and helping to shape the destiny of the political community" (Sandel, 1996: 5). This public philosophy views citizens as having a sense of the common good and a moral bond with their neighbours. Stemming from civic republicanism, this public philosophy is referred to as deliberative democracy because it builds on "a politics of engagement" (Kemmis, 1990: 12).

It depended first upon people being deeply engaged with one another ("rejoicing and mourning, laboring and suffering together") and second upon citizens being directly and profoundly engaged with working out the solutions to public problems, by formulating and enacting the "common good" (Kemmis, 1990: 12).

Deliberative democratic and competitive pluralist public philosophies are compared and contrasted in Table 12.1 by using a framework developed from work by Stanley (1988). This framework compares each public philosophy on four points: education outcome, consensus achieved, an experiential analogy, and role of participants.

Table 12.1. Comparison of two world philosophies (adapted from Stanley, 1988; and Sandes, 1996).

Result of Public Philosophy	Public Philosophy	
	Competitive Pluralism	Deliberative Democracy
Education Outcome	Educate people in range of policies constrained by market economy.	Civic education, experiential inquiry, social learning, treats citizens as policymakers
Consensus Achieved	Evaluate trade-offs among options deleloped by experts.	Shared, ongoing narrative; civic conversation re: where we are and where we want to go - defines common vision; civic engagement; individual and group identity.
Experiential Analogy	Grief-coping, focus on being deprived of something.	Immigration - working through change to transcend and adapt to achieve improvement.
Role of Participants	Aggregates of individuals with preformed wants, and values based on individual interests.	Complex social beings, inherent obligation to transcend and adapt to achieve improvement.

4.1 Education Outcome

Deliberative democracy encourages experiential inquiry and social learning. As defined by Korten (1981: 614), social learning involves simultaneous generation of "new knowledge, new benefits, and new action potentials as integral outcomes of a single process." In contrast, competitive pluralism seeks to educate people about policies developed by experts while basing decisions and the development of policies on the economics of market demand.

Consensus achieved. A deliberative democratic public philosophy strives to achieve a shared, cumulative story (Stanley, 1988). Deliberative democracy is not about difficult tradeoffs, or creating winners and losers. Instead the focus is on broadening the understanding of issues through experiential inquiry. Thus, "consensus" is achieved through ongoing public narrative. This concept of consensus differs from the one implied in the competitive pluralist philosophy, which requires evaluating tradeoffs among pre-developed alternatives.

4.2 Experiential Analogy

Stanley (1988) develops an experiential analogy for each public philosophy. Although he sees competitive pluralism resulting in a process similar to grieving, democratic deliberation, he says, is more like immigration in that people are able to transcend change and a feeling of loss. There is an emphasis on sharing one's experiences while working through change. Transcending change in this way builds on ideas of civic friendship and empathy for the situation of our neighbours.

4.3 Participants

A deliberative democracy conceives of participants as "complex social beings" whose perceptions are influenced by groups and individual affiliations, past experiences, and other social and psychological factors (Stanley, 1988: 14). Rather than atomized persons with no obligations beyond their own self-interest, participants have obligations based on their "encumbered identities" as members of various social worlds including families, cultures, racial and ethnic traditions, neighbourhoods, voluntary associations, and professions.

Based on these differences, the two public philosophies differ greatly in their approaches to political participation. Because these public philosophies inform theory and method, they have implications for social science research, resource planning, and other resource management practices. From a competitive pluralist perspective, political participation by the public is often seen as chaotic and unnecessary. Public participation is frequently reduced to lobbying, referenda or lawsuits to settle contentious issues. In resource management and planning, competitive pluralism is evidenced by choosing a method that relies on expert scientific analysis and policy elites for their ability to make "rational" decisions (Stanley, 1988). This method is what Bryan (1996: 149) refers to as a "technocratic approach". Within this approach, "data are considered suspect unless gathered by experts, analyzed by experts, and interpreted to the public by experts" (Bryan, 1996: 149). Bryan (1996: 149) notes, "Implicit in this orientation is that the public has neither the skills nor the responsibility to be actively involved in the process of planning."

Within a philosophy of competitive pluralism, processes are put into place that avoid bringing parties with differing viewpoints together to deliberate on a common good. Formal public hearings allow public comment but no discussion. Even open houses have limited opportunities for conversation among citizens. Few formal public involvement processes facilitate what Yankelovich (1991) refers to as a "working through" process. Instead of processes that bring citizens together to formulate values and interests through conversation, formal processes such as the methods of standard social assessment and social impact assessment for "weighing, balancing, and upholding rights" have been institutionalized (Kemmis, 1990).

In contrast, a deliberative democratic public philosophy is built on ideas of civic humanism (Stanley, 1988). A deliberative democracy values democratic participation and civic engagement and builds on a conception of shared values, or common ground (Kemmis, 1990) that emerges through the deliberative process.

Thus, within this philosophical orientation, there is an effort to provide forums for people to come together as citizens, to deliberate about themselves and the common good (Stanley, 1988).

In this view, citizens are conceived of as policymakers interested in understanding issues and implications of alternative courses of action. Citizens create knowledge in deliberative processes that include observation, interpretation, and involvement (Shannon and Antypas, 1996). "Knowledge production is integrated with and therefore cannot be separated from the enlightenment function of self-discovery and the moral effects of political deliberation and choice" (Shannon and Antypas, 1996: 68).

Within competitive pluralism, there are few formal or institutionalized opportunities for face-to-face democratic participation and collaborative activities. A deliberative democratic public philosophy, on the other hand, embraces political participation and civic engagement (*e.g.* social learning processes) as ends in themselves by which "to rediscover the civic commons and its associated identity of citizenship" (Stanley, 1988: 8).

5   Social Learning and Collaboration

Social learning provides a framework for (a) sharing understanding of the context, uniqueness, and importance of a place; (b) exploring issues, concerns, solutions, and management strategies; and (c) understanding judgments of acceptability. Social learning theory comes from the perspective that people's preferences are formed through the very processes designed to discover and respond to them. In other words, we do not have a preference for something until we actually think about making a judgement about it. From a social learning perspective, it does not make sense to hold a public hearing to solicit people's perspectives unless they also are provided with opportunities to "work through" (Yankelovich, 1991) the issues to be commented on. Based on social learning theory, the critical quality of planning and decision making processes is to facilitate learning - both public and organizational learning for all participants.

Problem definition and solution generation are meaningful social learning processes. We humans like to figure things out for ourselves. We tend to place more value on something we have been part of discovering for ourselves than on something someone else - especially someone in a position of power or authority - tries to "educate" into us. Learning by doing is much more powerful than being told something. The challenge for resource agencies is to design processes that facilitate opportunities for "working through" (Yankelovich, 1991) such that judgements become shaped through social interaction, deliberation, and learning.

As previously mentioned, a variety of stakeholders must be engaged in determining what to sustain and what indicators adequately measure success (McCool and Stankey, 1988). Civic engagement in deliberation, decision making, stewardship activities, civic science, and other processes that bring people together to explore options and potential outcomes is needed. Fortunately, citizens worldwide are becoming increasingly interested in participating in planning, decision making, and stewardship activities. Facilitating opportunities for meaningful citizen participation can improve the possibility of achieving sustainable forest management (Shindler and Cramer, 1999). Collaboration is recognized as a demonstration of democracy, a way to ensure that citizens are heard, a way to make sure that management is acceptable to citizens, and as a way to give legitimacy to processes by incorporating local knowledge and values and opening up expert knowledge (and opinion) to public scrutiny.

The recognition and acceptance of the dual nature of participation leads to the adoption of an attitude that both presupposes and anticipates (a) public responsibility for environmental sustainability and social justice; (b) the social conditions necessary for participatory planning and design, as those which allow for the integration of administrative levels, segments of society, subject domains, and localities across space; and (c) a reflective, critical planning and design process that will result in the transformation of both the social conditions and the knowledge bases, thus contributing to both understanding and building sustainable places (Sancar, 1994: 333).

Rather than using a reflective, critical process that facilitates participatory planning, resource agencies often approach landscape management by using tools consistent with a technical, rational model to assess visual preference for one landscape over another (*e.g.* scenic beauty estimation). Although it is interesting to explore which landscapes people prefer, this information is not useful in helping us understand how people encounter and interact with the landscape, the meanings they ascribe to it, what their expectations of management are, or what those expectations are based on.

Thayer (1989) suggests that people who understand how and why a landscape functions, taking into account biophysical and social and political influences, will have different responses to sustainable management practices than those who do not understand how a landscape functions. If this is true, then we each will respond to a landscape differently depending on our level of understanding. This presents an interesting research question: What factors contribute to how we respond to a landscape? One might hypothesize that while participation with a group on a field trip might not change someone's preferences, it might result in a different level of acceptability of a particular management practice or condition as people gain a broadened perspective through dialogue with others.

Facilitating open discussions about sustainable landscapes will enable people to become more conversant about what they are seeing, what the intent of management is and why a particular action was chosen or decision made. Increased participation in the conversation will help the public come to know and possibly accept sustainable management activities (Thayer, 1989). From this we could also suggest that discussions around unsustainable landscapes and what it is that leads us to define them as unsustainable might help the public redefine what it is they want to sustain, reject unsustainable activities and support sustainable ones based on their redefinition. "Human beings are symbolic animals, and it is through the relation of form, structure, process, and information content that sustainable landscapes will acquire meaning and provide depth of experience" (Thayer, 1989: 109).

Research has found that judgments of acceptability are based on more than aesthetic preference (Brunson, 1996; Stankey, 1996; Shindler and Cramer, 1999). Several factors have been found to enter into a judgement of perceived landscape acceptability. These factors include, but are not limited to, what we see and how we interpret what we see, our attitude toward the environment and resource management, our personal experience, our knowledge of and attachment to the place, and what we think of the decision making process.

Sustainable practices can be promoted through access to knowledge and social learning processes. Shindler and Cramer (1999) suggest that unless attention is paid to issues of adequate forums for deliberation, credibility, trust, and confidence in the decision making process, and questions of risk and uncertainty, even high-quality scientific knowledge will not achieve or sustain public support or acceptability.

6 Conclusions

Social learning is a critical component of sustainability. As previously mentioned, perceived conflicts between sustainable forests and visual quality might have more to do with judgements of perceived appropriateness of the planning process; perceived intent of the management activity; familiarity or past experience with the place, process, or agency; trust in the agency; and several other personal and socio-cultural aspects than with aesthetic value. We may not be able to separate the influences of aesthetic value from these other factors; however research has found that how people base judgements of acceptability can be explored and modified through social learning processes.

The pursuit of sustainability requires attention to relations and processes as well as desired outcomes. Sustainability entails attention to the integration of interdependent ecological, cultural, social, and economic factors. Implementation of sustainability will be unique to particular places. Also unique to particular places are the relations people have with these places. How does a community's identification with a place and how it defines itself in relation to a place affect decisions about acceptable management? These relations play an important role in efforts to achieve sustainability. What are the relations between aesthetics and other aspects of place that people value? This is an area where we lack understanding and more research is needed. For example, aesthetic judgments are often proxies for other place-based values people hold. Environmental groups may capitalize on images of clearcuts to evoke reactions that go beyond the visual image itself to multiple aspects of what the clearcut represents. What are the underlying aspects and how can we better understand the values evoking the reactions?

What we do know is that people care about places, they have knowledge that is beneficial to decision making, and they must be included in decision making processes. What we need to learn more about is how local knowledge of places can be incorporated with scientific knowledge to help people make informed decisions. The institutions and forums necessary to enable adequate participation are lacking (Kusel and Fortmann, 1991; Shannon, 1991; Krannich *et al.*, 1994).

Current trends include an interest in preserving and sustaining biophysical landscapes and species with a separate and parallel interest in sustaining social communities. Human communities are part of landscapes located in place and time. What is lacking is a discussion of humans as a species embedded in landscapes. We need frameworks that can help us understand relations between human communities and the larger ecosystems of which they are a part. Examples of such frameworks may be found in some of the Adaptive Management Area projects undertaken by the US Forest Service (Clark *et al.*, 1998) and particularly the White Pass Discovery Team and Community Self-Assessment processes (Kruger, 1996). The Discovery Team project in particular brought together community members, school students, resource managers and scientists to learn together about the relations between the community and the encompassing ecosystem. Projects like this can help bring together biophysical and socioeconomic aspects of landscape and further efforts toward sustainability of human communities and biophysical and social aspects of landscapes.

References

Brunson, M.W. (1996) A definition of "social acceptability". In: M.W. Brunson, L.E. Kruger, C.B. Tyler and S.A. Schroeder, (tech. eds) *Ecosystem Management. Defining Social Acceptability in Ecosystem Management: A Workshop Proceedings*; 1992 June 23-25, Kelso, WA. Gen. Tech. Rep. PNW-GTR-369. US Department of Agriculture, Forest Service, Pacific Northwest Research Station, Portland, OR, pp. 7-16.

Bryan, H. (1996) The assessment of social impacts. In: A.W. Ewert (ed) *Natural Resource Management*. Westview Press, Inc., Boulder.

Clark, R.N., E.E. Meidinger, G. Miller, J. Rayner, M. Layseca, S. Monreal, J. Fernandez, and M.A. Shannon (1998) *Integrating Science And Policy In Natural Resource Management: Lessons And Opportunities From North America*. Gen. Tech. Rep. PNW-GTR-441. US Department of Agriculture, Forest Service, Pacific Northwest Research Station, Portland, OR.

Daniel, T.C. and R.S. Boster (1976) *Measuring Landscape Esthetics: The Scenic Beauty Estimation Method*. Res. Pap. RM-167. US Department of Agriculture, Forst Service, Rocky Mountain Forest and Range Experiment Station, Fort Collins, CO.

Dixon, J.A. and L.A. Fallon (1989) The concept of sustainability: Origins, extensions, and usefulness for policy. *Society and Natural Resources* 2: 73-84.

Fishwick, L. and J. Vining (1992) Toward a phenomenology of recreation place. *Journal of Environmental Psychology* 12: 57-63.

Forest Ecosystem Management Assessment Team (FEMAT) (1993) *Forest Ecosystem Management: An Ecological, Economic, and Social Assessment.* US Department of the Interior, Portland, OR. [Irregular pagination].

Gale, R.P. and S.M. Cordray (1994) Making sense of sustainability: Nine answers to "What should be sustained?" *Rural Sociology* 59(2): 311-332.

Gobster, H. and R.E. Chenoweth (1989) The dimensions of aesthetic preference: A quantitative analysis. *Journal of Environmental Management* 29(1): 47-72.

Greider, T. and L. Garkovich (1994) Landscapes: the social construction of nature and the environment. *Rural Sociology* 59(1): 1-24.

Kemmis, D. (1990) *Community and the Politics of Place.* University of Oklahoma Press, Norman.

Korten, D.C. (1981) The management of social transformation. *Public Administration Review* (Nov/Dec): 609-618.

Krannich, R.S., M.S. Carroll, S.E. Daniels and G.B. Walker (1994) *Incorporating Social Assessment and Public Involvement Processes Into Ecosystem-Based Resource Management: Applications To The East Side Ecosystem Management Project.* Interior Columbia Basin Ecosystem Management Project, Walla Walla, WA.

Kruger, L.E. (1996) *Understanding Place as a Cultural System: Implications of Theory and Method.* Unpublished Ph.D. dissertation. University of Washington, Seattle.

Kusel, J. and L. Fortmann (1991) *Well-Being In Forest Dependent Communities: Volume 1.* Department of Forestry and Resource Management, University of California, Berkeley, CA.

McCool, S.F. and G. Stankey (1998) *Representing The Future: A Framework For Evaluating The Utility Of Indicators In The Search For Sustainable Forest Management.* A paper presented at the international conference on indicators for sustainable forest management; 1998 August 24-28; Melbourne, Victoria, Australia.

Moore, S.A. (1997) "Place" and sustainability: research opportunities and dilemmas. In: F. Vanclay and L. Mesiti (eds) *Sustainability and Social Research: Proceedings of the Conference of the Australian Association for Social Research*; Charles Sturt University, Wagga Wagga; 1997 February, Wagga Wagga, New South Wales: Centre for Rural Social Research, Charles Sturt University: 217-229.

Nassauer, J.I. (1983) Framing the landscape in photographic simulation. *Journal of Environmental Management* 17: 1-16.

Orions, G.H. (1990) Ecological concepts of sustainability. *Environment* 32(9): 10-15; 34-39.

Orr, D. (1992) *Ecological Literacy: Education and the Transition to a Postmodern World.* State University of New York Press, Albany.

Relph, E. (1976) *Place and Placelessness.* Pion, London.

Rodman, M.C. (1992) Empowering place: multilocality and multivocality. *American Anthropologist* 94 (3): 640-656.

Sagoff, M. (1992) Has nature a good of its own? In: R. Costanza, B. Norton and B.D. Haskell (eds) *Ecosystem Health: New Goals for Environmental Management*. Island Press, Washington, DC.

Saint-Exupery, A. (1943) *The Little Prince*. Harcourt, Brace and World, New York.

Sancar, F.H. (1994) Paradigms of postmodernity and implications for planning and design review processes. *Environment and Behavior*. 26(3): 312-337.

Sandel, M.J. (1984) Morality and the liberal ideal. *New Republic* 16.

Sandel, M.J. (1996) *Democracy's Discontent: America in Search of a Public Philosophy*. The Belknap Press of Harvard University Press, Cambridge, MA.

Shannon, M.A. (1991) *Building Public Decisions: Learning Through Planning. An Evaluation of The NFMA Forest Planning Process*. Washington DC: Office of Technology Assessment.

Shannon, M.A. and A.R. Antypas (1996) Civic science is democracy in action. *Northwest Science*. 70(1): 66-69.

Shindler, B. and L.A. Cramer (1999) Shifting public values for forest management: Making sense of wicked problems. *Western Journal of Forestry* 14(1): 1-7.

Shumway, N. (1991) *The Invention Of Argentina*. University of California Press, Berkeley.

Social Science Research Group (1994) *Principles of Sustainability*. Unpublished report prepared by the Social Science Research Group, Institute for Resources in Society, College of Forest Resources, University of Washington, Seattle.

Stankey, G.H. (1996) Defining the social acceptability of forest management practices and conditions: Integrating science and social choice. In: M.W. Brunson, L.E. Kruger, C.B. Tyler and S.A. Schroeder, technical editors. *Defining Social Acceptability in Ecosystem Management: A Workshop Proceedings;* 1992 June 23-25, Kelso, WA. Gen. Tech. Rep. PNW-GTR-369. US Department of Agriculture, Forest Service, Pacific Northwest Research Station, Portland, OR., 99-111.

Stanley, M. (1988) The rhetoric of the commons: Forum discourse in politics and society. *Teachers College Record* 90(1): 1-15.

Thayer, R.L. (1989) The experience of sustainable landscapes. *Landscape Journal* 8 :101-110.

Unger, R.M. (1984) *Knowledge and Politics*. The Free Press, New York.

Walter. E.V. (1988) *Placeways: A Theory of the Human Environment*. The University of North Carolina Press, Chapel Hill.

World Commission on Environment and Development. (1987) *Our Common Future*. Oxford University Press, Oxford.

Yankelovich, D. (1991) *Coming To Public Judgment: Making Democracy Work In A Complex World*. Syracuse University, Syracuse.

PART V

Visualization of
Forested Landscapes

Chapter thirteen:

The Rhetoric of Visual Simulation in Forest Design: Some Research Directions

Don Luymes
Landscape Architecture Programme and Department of Forest Resources Management, University of British Columbia
Vancouver, British Columbia

1   Introduction

In recent years, there have been impressive advances in the ways in which landscapes - and landscape changes - can be simulated. These advances are due to rapidly advancing graphical computer technology that allows digitally-produced photorealistic and virtual reality simulations of both existing and imagined landscapes (Oh, 1994; Orland and Uusitalo, Chapter 14 this volume; Danahy, Chapter 15 this volume). Such advanced simulations of landscapes have obvious benefits. They can effectively demonstrate the visible effects of land management decisions and strategies before they occur, in images that are readily understood by most people. Alternative land management strategies can be quickly and realistically compared; several views of the same landscape - taken from different viewpoints - can be viewed in succession or simultaneously on a computer screen or projected onto a wall. Changes in the composition of a landscape over time can be shown. In the most advanced, virtual reality simulations, a viewer can negotiate her way into and through a seemingly three-dimensional landscape, controlling her direction and speed of travel, her field of view and the objects upon which she chooses to focus.

The development of such simulation tools is directly relevant to forest management, particularly the management of public forests. Public involvement in forest land decisions has become expected in North America, and accessible simulations can illustrate the implications of alternative forest plans, allowing more meaningful public involvement in forest management. However, for all of the potential (and very real) benefits of advanced, computer-generated landscape simulations, there are also reasons for caution, and there are problems that are particular to the most sophisticated simulation technologies.

In this chapter I will briefly outline the benefits of photorealistic and virtual reality landscape simulations, with particular reference to the role of visual simulations in the design and management of forest landscapes. I will also outline some inherent problems that may be particularly associated with photorealism and state-of-the-art visualization technology. I will then outline a suggested agenda for research in this dynamic, interdisciplinary field. In summary I will argue that simulations are rarely - if ever - value-neutral, but are instead rhetorical devices (in the Aristotelian tradition of *means of persuasion*). As such, simulations can be effectively mobilized in the making of more sustainable landscapes that are also perceived as beautiful by the public.

2 The Power of Imagery

In a society that has been characterized as *hyper-representational* (Harvey, 1989), where the visual image is increasingly important and pervasive it comes as no surprise that visual simulation has taken on an increasingly central role in environmental decision-making. Terry Daniel has noted that "it is impossible to imagine any significant natural resource management activity that does not rely to some extent on visual representations" (Daniel, 1992: 261). Over the past few decades, members of the public have become increasingly concerned with the visible effects of environmental change, particularly in sensitive landscapes (Sheppard, 1989). Lay persons, or the general public, are demanding involvement in forest and wildland decisions that were traditionally made by some class of professionally-educated experts or by society's elite. In this milieu, visual images of how change will appear are the common currency; unlike technical language and data, visual images are easily readable and understood by the public.

Concurrent with the demands for public scrutiny, and public involvement in landscape planning, advances in computerized visualization technology now allow a high degree of *perceived realism* (Perkins, 1992) in visual landscape simulations. This high level of perceived realism presumably results in a greater degree of agreement between peoples' aesthetic reaction to simulations and the real environment that is being represented (Bishop and Leahy, 1989; Sheppard, 1989; Bosselmann, 1998).

The effects of these trends are that visual imagery is used as a kind of talisman in debates over ecological issues, notably those surrounding forest management and timber harvesting. Two examples will illustrate this point. In 1993 a glossy, full colour large-format book was published by the Sierra Club, entitled *Clearcut*. Unlike most books of this coffee-table genre, this book features scenes of destruction and devastation, rather than scenes of beautiful, undisturbed natural landscapes (Devall, 1993). Using powerful visual imagery of recent, large cutblocks, landslides and degraded streams, the book's intention is to argue that the ecological health of managed forest lands in North America is in critical condition. This visible evidence is offered as a potent sign, signifying and implying ecological and spiritual damages that are in fact largely unseen.

In opposition, the Working Forest Project, (an initiative of a group of professional foresters, most from private industry) also offers visual images as evidence of ecological conditions in the book *The Working Forest of British Columbia* (Robson, 1995). However, in this case, before and after photographs of rapid green-up on logged and replanted sites are shown as a proof of ecological resiliency and biological productivity. The implication is that since the visible scars of logging are covered by vigorous regeneration, the forest itself remains healthy.

Both of these books use visual, photographic evidence (which is, of course, itself a simulation) to support ecological and political arguments, regardless of whether the images directly address the ecological questions, or whether the simulations offer a complete picture (Kimmins, Chapter 4 this volume). In a similar vein, as advances in graphical computing power and speed continue, and as simulations become increasingly sophisticated, it is likely that visual simulation of future conditions will become even more central in debates and discussions over environmental sustainability.

3 New Areas of Advance

New *arenas of representation* have been opened up in recent years, based on computing power, speed, and graphic capabilities that have become more readily available, less costly and cumbersome to use in practice (Lange, 1994; Orland and Uusitalo, Chapter 14 this volume; Danahy, Chapter 15 this volume). Three of these new arenas are highly realistic (virtual reality) three-dimensional environments, detailed simulation of forest growth and change over time and reconstructed digital photographs.

Digital modelling is the loose term used to cover a variety of techniques that involve the construction of three-dimensional, scaleable digital environments in accurate perspective. These systems can display landscape features that are accurate in scale and spatial relationship, with detailed, textured surfaces. More advanced systems introduce real-time, viewer-controlled simulated movement through three-dimensional landscapes; by adding sound (and even smells and tactile pressure) the viewer is given the sensation of bodily entry into the simulated environment. This technology is called virtual reality (VR); similar technology has been developed in the computer animation industry and is employed in flight simulation. Relatively simple VR is now finding its way into environmental simulation.

A second arena of simulation advance involves the development of fractal algorithms that simulate vegetation growth and change in three dimensions. This technology differs from conventional computerized simulation in that the spatial forms and patterns emerge in a growth-like fashion - from the inside, seeming to mimic the process as well as the patterns of vegetative growth and distribution (McQuillan, 1995).

A third, major arena of computer-driven simulation is digital imaging. Photorealistic digital image manipulation is possible using relatively inexpensive, widely available technology. Using off-the-shelf software programs, illustrators can manipulate and modify digital images that are either captured by a digital camera or scanned from slides or print photographs (Orland, 1994). If the illustrator is skillful, the resulting modified images are indistinguishable from photographs of real landscapes. If the viewpoint locations and directions of view are calibrated to an accurate digital terrain model in a Geographic Information System (GIS), a high level of accuracy can also be attained (Lange, 1994; Orland, 1994).

4 Some Benefits of Computer-Based, Photorealistic Simulation

The expected benefits of these photorealistic or virtual reality simulations are based on a simple, three step logical progression. First, by definition, photorealistic computer simulations have a high degree of perceived realism and public recognition. These simulations are often so lifelike that they are mistaken for photographs of existing places (Sheppard, 1989). Due to our society's familiarity with photographic and videographic conventions and imagery, their subject contents (landscape features and conditions) are easily recognized and interpreted. Research has shown that photographs (Shuttleworth, 1980) and digitally doctored or constructed photographs are an adequate surrogate for real environments, eliciting similar aesthetic responses from viewers (Bishop and Leahy, 1989; Hull and Stewart, 1992; Bergen *et al.*, 1995). So, for example, if one desires to simulate a mature forest, it can be expected that certain patterns of green, brown and grey colour pixels on a computer screen or dots on a page will be correctly interpreted as such.

Second, this high degree of public recognition has the potential effect of creating a more widely understood *language of discourse* around issues of environmental change (Bosselmann, 1998). Since this information is packaged in a highly recognized form (the photographic image) that does not need expert interpretation, implications embedded in the simulation are also recognized and understood. For example, reddish foliage digitally painted onto the previously-mentioned forested scene can be expected to be interpreted correctly as a spatially-extensive disturbance due to insect damage or fire.

Third, better access to information about environmental changes will presumably lead to more informed, or more widely supported decisions about those changes, based on their visible effects (Lange, 1994). However, it must be acknowledged that visible clues or effects of environmental change may not give solid evidence of ecological health, and may sometimes mislead the public about ecological effects. Following our previous example, it can be expected that a photorealistic simulation of an insect outbreak will prompt an informed (and possibly rancorous) public debate on the relative merits of alternative forest pest management strategies - whether or not the ecological ramifications are fully understood.

Value Judgements

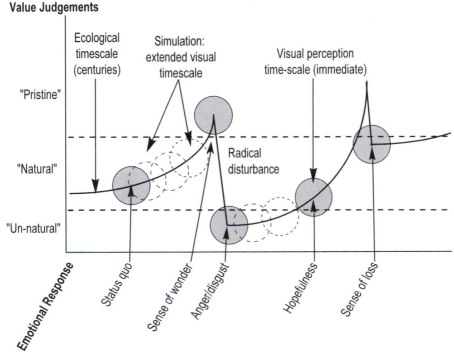

Fig. 13.1. Ecological and visual time-scales. The potential for simulation of forest conditions to resolve the dilemma between long ecological time-scales and the immediate visual apprehension of forest scenery.

Virtual reality (VR) simulations add three other factors to the ones mentioned above. VR introduces the dimension of time and movement into the simulation, increasing the amount of visual information available to the viewer, and presumably offering a more holistic, detailed sense of the simulated landscape (Pitt and Nassauer, 1992; Orland and Uusitalo, Chapter 14 this volume). VR simulations that allow the viewer to choose their own path through the simulated environment provide even greater opportunities for exploration and cognitive mapping (Lynch, 1960). And, VR imagery can be tied to other sensory information (sounds, smells, even tactile sensation) that can serve to further immerse the viewer into the world of the simulation. However, it should be noted that the benefits of a viewers' experiences of VR are still captured in the three-step chain of benefits listed above.

A distinct, additional benefit of powerful simulations that include the modelling and display of vegetative growth and landscape change is their ability to break down the time-scale dilemma between ecological process and visual apprehension (Kimmins, Chapter 4 this volume). This dilemma emerges because many ecological processes (like fire disturbance regimes, vegetative succession, hydrological and geomorphological processes) take many years or centuries to unfold. However, a visual apprehension of these processes (whether viewed in person or through a photograph) is usually time-specific, a snapshot of only one moment in the process (Figure 13.1). As a result, an observer's value judgments and emotional responses to particular landscapes are based on the present successional condition of that landscape. Sophisticated simulations can effectively and realistically portray changing landscape conditions over time, so that a viewer can attain a broader perspective, a glimpse (or a projection) into the future.

5 Problems of Simulation Content and Interpretation

While the benefits of advanced photorealistic and virtual reality simulations of landscapes and landscape changes are real, I would argue that along with these benefits come some potential problems. These can be classed as *problems of simulation content* and *problems of simulation interpretation.*

5.1 Problems of Simulation Content

Problems of simulation content emerge when there is divergence between the landscape simulation and the real world data that supports the simulation, or when the simulation is demonstrably inaccurate in its portrayal of visible real world phenomena. These problems of content can be found in simulations of present landscape conditions as well as future landscape conditions.

In the present, for example, the structural and floristic complexity of forested landscapes defy (or frustrate) large-scale, highly specific three-dimensional simulation (Orland and Uusitalo, Chapter 14 this volume). Even the best, highest resolution VR simulations, with texture maps draped over detailed 3-D terrain models, are edited, simplified forests; the cost of creating and displaying realistic detail of something as structurally complex as a multi-layered temperate rainforest is probably prohibitive in the foreseeable future.

As a way of realistically portraying three-dimensional detail, a sort of visual shorthand is employed. Classes of similar objects (like individual trees) are not modeled separately, but are repeated. Typically there is a structural simplification of the forest scene. For example, this may involve the removal of understorey layers or the filling in of surfaces like ground planes and foliage that fall between the mesh of sampled data points. Of course, such editing and artistic license is probably inevitable, but the extent to which the simulation diverges from the data that supports it is conceptually problematic.

In simulation of future landscapes, problems of content can also result from the inherent uncertainty of future forest conditions, due to variables like growth rates, interspecies competition, changing climatic conditions, fire and wind-induced disturbance, insect and parasitic infestations. All of these stochastic processes are largely unpredictable over the kind of timeframes that are necessary to simulate forest development, and even slight variations in the modeled conditions will result in widely divergent visual effects. In short, these simulations have a lower level of confidence than simulations of present-day landscapes. Of course, it must be noted that this is true of all simulations, not just computer-driven ones. In fact, it could be argued that computer-generated simulations are less capricious and subjective than previous means of simulation (Sheppard, 1989; Daniel, 1992; Bosselmann, 1998).

However, simulations are also subject to problems of interpretation; no matter how accurate the simulation, it does not guarantee accuracy of understanding by those who are viewing the simulation.

5.2 Problems of Interpretation: Assumptions of Transparency and Authority

Problems of interpretation are related to the ways in which people read, or interpret simulations. These problems emerge when viewers carry assumptions that confuse (or misinterpret) the true relationship between the simulation and the real environment it is portraying, between the sign and that which it signifies (Steinitz, 1992). Photorealistic simulations heighten these problems, since the public is more likely to assume reality when faced with a high level of perceived realism. The assumptions that lead to problems of interpretation include *assumptions of transparency* and *assumptions of authority* (Figure 13.2).

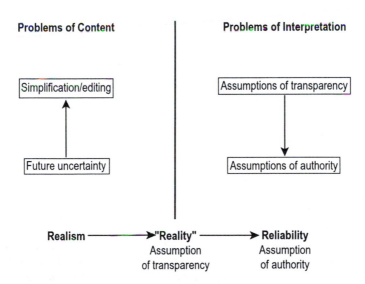

Fig. 13.2. Problems of interpretation that are particularly acute for photorealistic or virtual reality simulations.

Assumptions of transparency are viewers' assumptions that a particular simulation is an accurate, *transparent* depiction of the real environment, even when people realize that the simulation is not a photograph of an existing landscape (Pitt and Nassauer, 1992). Assumptions of authority are based upon the assumption of transparency: once there is a belief in accuracy, the simulation (and its creators) are considered authoritative - they are trusted to be a reliable predictor of future conditions. The sequence of interpretation of such simulations can thus be conceived of as a logical progression from *realism* to *reality* to *reliability* (Figure 13.2).

Simulations that show with purported - or implied - confidence what a forest will look like given certain conditions or silvicultural treatments, run the risk of overstepping the level of confidence or detail that their data inputs permit. While professionals may understand this discrepancy, and while disclaimers may be attached to the simulation, the public may not be able to separate the high levels of realism embodied in photorealistic simulations from uncertainty about their level of accuracy. The very power and potency of photorealistic simulations can create this unintended consequence.

As simulations become - and more importantly *purport* to become - more lifelike, more realistic the potential for both illuminating and obscuring motives, limitations and assumptions increases. If the *transparency* of photographic imagery is assumed, *authority* is more readily granted or assumed as well, and there is less room for skepticism or for questioning the assumptions that the simulation is based upon. This condition may be compounded by a certain awe of high technology in the public, also resulting in less critical attitudes towards the medium and its message. Problems of interpretation become even more serious, of course, when a simulation is intended to deliberately mislead. In this case, the sophistication of the technology itself may heighten the risk that disingenuous simulations are used for overt (or covert!) political purposes.

6   Addressing the Problems

To overcome problems of content and interpretation, there is first of all a simple need to acknowledge them - both internally (among simulation researchers and practitioners) and externally (to the public, and to clients).

In addition, there is a need for diverse, multi-media simulations that illustrate the range of possibilities of a given management regime. Such representations should use different media (wire-frame computer drawings as well as draped texture surfaces; hand-drawn diagrams and numerical data as well as photorealistic digital imagery) at different scales and levels of resolution and abstraction. Such diverse representations of a variety of types accomplish two objectives: illustrating to the public that the future look of the landscape is less certain than a single representation would imply, and eliciting a more fully-rounded public understanding of the project or management strategy that is being contemplated (Daniel, 1992).

There is also the need for full disclosure (*open* simulations) that indicate the degree of uncertainty that certain visible effects will occur, and that indicate the degree of editing of content that was necessary to construct the simulation (Orland and Uusitalo, Chapter 14 this volume). Finally, there is a need for visual simulations to be explicitly tied to defensible spatial and ecological data, so that the unconscious tendency to compose and frame scenes (what Gina Crandell calls *pictorialization*: 1993) is reduced.

7 Agendas for Research

If the benefits of better, more accurate visual simulation of forest landscapes are to be realized, while guarding against the problems inherent in the content and interpretation of such realistic images, four broad classes of research should be carried on simultaneously. These research tracks might be called *display research*, *media research*, *control research* and *process research*.

7.1 Display Research

Display research involves the technical pursuit of creating more accurate simulations of natural environments, including ways of more efficiently and accurately modeling the complexities of the forest. Such complex simulations will include snags and old growth-like trees, woody debris and windfall, understorey shrubs and second-level canopies. One promising technique to refine is the simultaneous projection of a subject landscape at two levels of resolution: life-like detail of representative, small-scale portions of the forest embedded within coarser, more abstract model environments. A second trail to follow is the projection and display of forest growth and change under different conditions and trajectories, from enhanced growth through low-level and radical disturbance regimes of different types.

7.2 Media Research

This second class of investigation is related to the previous one. In order to address the reliability and defensibility of simulations, more effective ways of combining numerical environmental data and visual display are needed (Daniel, 1992; Steinitz, 1992; Orland and Uusitalo, Chapter 14 this volume). One possible avenue of research is in developing more seamless connections, so that if a variable is changed in the supporting data (for instance, climate, or growth rates) the corresponding visual information encoded in the simulation also changes. Also, new ways of presenting and even building simulations via the medium of the Internet offer promising directions for further investigation.

7.3 Control Research

This research track involves ground-truthing, validity testing and establishing complexity thresholds. How do simulations compare with control landscapes, and what is the congruence between peoples' responses to various simulation media and directly-experienced landscapes? There is an established stream of this research that needs to be continued and widened to include new types of simulation, like VR (Sheppard, 1983 and 1989; Bishop and Leahy, 1989; Bishop, 1997).

Also, how much detail or how much realism is enough to adequately convey the intent and extent of management decisions (Perkins, 1992)? How much detail or realism in a simulation is too much - so that *cognitive perspective* (the ability of the viewer to retain a critical distance) is reduced (Steinitz, 1992)? How can structurally complex environments (forests with old-growth characteristics, for example) be most adequately simulated? What components of the simulation (colour, detail, context, motion, scale) influence perceived realism or mask the true nature of the content (Sheppard, 1989; Perkins, 1992; Bishop, 1997)?

7.4 Process Research

This final class differs from the first three that have been discussed: its emphasis is on the cultural production of simulations, on the motivations of their sponsors and on social effects of their uses. Potential questions include how simulations are (or could be) used in public decision-making, how various levels of sophistication in simulations influence public responses and on how simulations are (and can be) used as persuasive and political tools. This last category is a critical, *soft* science and humanistic scholarship, and it is imperative that there be communication between the scholarship produced in this group and the technically-based research that drives the other three groups.

8 The Rhetorical Use of Simulations For a More Sustainable Forestry

I will conclude by arguing that simulations are inevitably rhetorical products, "assertions made for the purpose of persuading some audience" (McClosky, 1985: 47). This is not to say that such rhetorical persuasion is sinister or intentionally misleading; it is merely a recognition that simulations (like any other documents) are constructed with an audience in mind, with a specific intention, and from within a framework of personal or agency values.

For example, a timber harvesting company may wish to demonstrate to the public that its plan for selective heli-logging on a steep slope will be visually unobtrusive and is designed to minimize the risk of landslides. Simulations of this intervention may be honestly intended and constructed as carefully and accurately as possible. In fact, the simulation construction may have been contracted out to a separate consulting firm or to a university research group; there may even be several alternative simulations that indicate a range of possible visual results. But such a simulation is obviously rhetorical, intended to persuade approval agencies and members of the public that the harvesting plan is reasonable and responsible.

Rather than seeing the rhetoric inherent in simulation as a failing, I believe that it is important to acknowledge the values that lie behind simulations, and to expose the rhetorical nature of simulations that purport to be value-neutral. Further (and here I declare my own bias), I believe that the rhetorical nature of visual simulation may be *openly* employed in the service of a more sustainable forestry - as that goal is currently understood.

As Carlson points out, public aesthetic preference is at least partly (and perhaps significantly) culturally determined (1993). When faced with forest harvesting - especially clearcuts - people's aesthetic responses are typically very low (BC Ministry of Forests, 1996), much lower than if they are convinced that they are looking at a naturally-caused disturbance, or even a pasture (Hodgson and Thayer, 1980; Moore, 1995). In the parlance of Levi-Strauss (1969), it is often the *cooked* interpretation (not the *raw* content) that determines much of the aesthetic response to particular, managed landscapes (Steinitz, 1990).

As several researchers have pointed out (Gobster, 1995; Sheppard, Chapter 11 this volume), not all of the physical, visible results of sustainable forestry (or *new forestry*) are congruent with landscapes that are conventionally judged to be beautiful or visually preferred. Examples of *new forestry* practices include leaving woody debris on logged-over sites and in riparian corridors; leaving or erecting snags in cutover areas; concentrating cutblocks rather than scattering them across a forested matrix; using prescribed burning or allowing lightning-induced fire to burn a forest (Kohm and Franklin, 1997). None of these practices yield landscape aesthetics that have been found to be preferred by the public (Schroeder and Daniel, 1980; Kaplan and Kaplan, 1989; Brown, 1994).

However, in a fascinating study that used photorealistic simulations to explore public aesthetic response to similarly messy but sustainable landscapes, Joan Nassauer found that evidence of care (*cues to care*) or of serious ecological intent caused aesthetic preferences for messy ecosystems to rise significantly (Nassauer, 1995). In a similar vein, several speculative and analytical papers (Koh, 1988; Spirn, 1988) have indicated or described a new ecological aesthetic - presumably indicating a shift in public aesthetic preferences that align to shifts in public ethical preferences (Carlson, Chapter 3 this volume). In other words, if the public knew what sustainable forestry practices looked like, they would be more likely to align an initially negative aesthetic response (*that's ugly*) with their ethical preference for sustainable practices (*I can see a certain beauty in that*).

It follows that simulations that are tied to contemporary scientific understanding, illustrating the visible effects of a more sustainable forestry practice can be potentially powerful tools for *shaping* (rather than merely acceding to or reflecting) prevailing public values. This is squarely in the Aristotelian definition of rhetoric as persuasion. Of course, it is equally important to also acknowledge the uncertainties inherent in these simulations themselves, and even the limits of current knowledge of what is sustainable.

This is a clear and important research agenda: simulation experts working closely with forest ecologists and foresters to develop time-sequenced simulations of forest succession under a range of sustainable management regimes. This may be the best glue of all between aesthetic preferences and more sustainable practices in forest management: using simulations to *influence* aesthetic preferences by illustrating sustainable practices, rather than merely looking (perhaps in vain) for areas of congruence between the two.

References

Bergen, R.D., C.A. Ulricht, J.L. Fridley and M.A. Gantor (1995) The validity of computer-generated graphic images of forest landscape. *Journal of Environmental Psychology* 15(2): 135-146.

Bishop, I.D. (1997) Testing perceived landscape colour differences using the Internet. *Landscape and Urban Planning* 37: 187-196.

Bishop, I.D. and P.N. Leahy (1989) Assessing the visual impact of development proposals: the validity of computer simulations. *Landscape Journal* 8: 92-100.

Bosselmann, P. (1998) *Representation of Places: Reality and Realism in City Design.* University of California Press, Berkeley.

Bourassa, S. (1991) *The Aesthetics of Landscape.* Bellhaven Press, London.

British Columbia Ministry of Forests (1996) *Clearcutting and Visual Quality: A Public Perception Study.* Ministry of Forests, Recreation Section, Victoria, BC.

Carlson, A. (1993) On the theoretical vacuum in landscape assessment. *Landscape Journal* 12 (1): 51-58.

Crandell, G. (1993) *Nature Pictorialized: "The View" in Landscape History.* Johns Hopkins Press, Baltimore.

Daniel, T.C. (1992) Data visualization for decision support in environmental management. *Landscape and Urban Planning* 21: 261-263.

Devall, B. (ed) (1993) *Clearcut: the Tragedy of Industrial Forestry.* Sierra Club Books/Earth Island Press, San Francisco.

Gobster, P.H. (1994) The aesthetic experience of sustainable forest ecosystems. In: W.W. Covington and L.F. DeBano (tech. coords.) *Sustainable Ecological Systems: Implementing an Ecological Approach to Land Management.* USDA Forest Service Gen. Tech. Rep. RM-24, pp. 246-255, Washington DC.

Gobster, P.H. (1995). Aldo Leopold's "ecological esthetic": integrating esthetics and biodiversity values. *Journal of Forestry* 93(2): 6-10.

Harvey, D. (1989) *The Condition of Post-modernity: an Enquiry into the Origins of Cultural Change.* Blackwell Press, Oxford.

Hodgson, R.W. and R.L. Thayer (1980) Implied human influence reduces landscape beauty. *Landscape Planning* 7: 171-179.

Hull, R.B. and W.P. Stewart (1992) Validity of photo-based scenic beauty judgments. *Journal of Environmental Psychology* 12: 101-114.

Kaplan, R. and S. Kaplan (1989) *The Experience of Nature: A Psychological Perspective*. Cambridge University Press, New York.

Koh, J. (1988) An ecological aesthetic. *Landscape Journal* 7 (2): 177-191.

Kohm, K. and J.F. Franklin (eds.) (1997) *Creating a Forestry for the 21st Century*. Island Press, Washington DC.

Lange, E. (1994) Integration of computerized visual simulation and visual assessment in environmental planning. *Landscape and Urban Planning* 30: 99-112.

Levi-Strauss, C. (1969) *The Raw and the Cooked*. Translated by J. Weightman and D. Weightman. Harper and Row, New York.

Lynch, K. (1960) *The Image of the City*. MIT Press, Cambridge.

McClosky, D. (1985) *The Rhetoric of Economics*. University of Wisconsin Press, Madison.

McQuillan, A. (1995) The potential of fractal geometry as a tool for achieving naturalism in forest aesthetics. In: K. O'Hara (ed), *Uneven-Aged Management Opportunities, Constraints and Methodologies*. MFCES Misc. Pub. 56, University of Montana School of Forestry, Missoula MT, pp. 51-56.

Moore, P. (1995) *Pacific Spirit: the Forest Reborn*. Terra Bella Publishers, Vancouver.

Nassauer, J.I. (1995) Messy ecosystems, orderly frames. *Landscape Journal* 14: 161-170.

Oh, K. (1994) A perceptual evaluation of computer-based landscape simulations. *Landscape and Urban Planning* 28: 201-216.

Orland, B. (1992) Data visualization in environmental management: a research, development and application plan. *Landscape and Urban Planning* 21: 241-244.

Orland, B. (1994) Visualization techniques for incorporation in forest planning geographic information systems. *Landscape and Urban Planning* 30: 83-97.

Perkins, N.H. (1992) Three questions on the use of photo-realistic simulations as real-world surrogates. *Landscape and Urban Planning* 21: 265-267.

Pitt, D.G. and J.I. Nassauer (1992) Virtual reality systems and research on the perception, simulation and presentation of environmental change. *Landscape and Urban Planning* 21: 269-271.

Robson, P. (1995) *The Working Forest of British Columbia*. Harbour Publishing, Madiera Park.

Schroeder, H. and T.C. Daniel (1980) Progress in predicting the perceived scenic beauty of forest landscapes. *Forest Science* 27: 71-80.

Sheppard, S.R.J. (1983) How credible are visual simulations? *Landscape Architecture* 73 (1): 83.

Sheppard, S.R.J. (1989) *Visual Simulation: A User's Guide for Architects,*

Engineers and Planners. Van Nostrand Reinhold, New York.

Shuttleworth, S. (1980) The use of photographs as an environmental presentation medium in landscape studies. *Journal of Environmental Management* 11: 61-76.

Spirn, A. W. (1988) The poetics of city and nature: towards a new aesthetic for urban design. *Landscape Journal* 7 (2): 108-126.

Steinitz, C. (1990) Toward a sustainable landscape with high visual preference and high ecological integrity: the Loop Road in Acadia National Park, USA. *Landscape and Urban Planning* 19: 213-250.

Steinitz, C. (1992) Some words of caution. *Landscape and Urban Planning* 21: 273-274.

Thayer, R.L. (1989) The experience of sustainable landscapes. *Landscape Journal* 8(2): 101-110.

Wood, D. (1988) Un-natural illusions: some words about visual resource management. *Landscape Journal* 7 (2): 192-205.

Chapter fourteen:

Immersion in a Virtual Forest - Some Implications

Brian Orland
Department of Landscape Architecture, Pennsylvania State University
University Park, Pennsylvania

Jori Uusitalo
Faculty of Forestry, University of Joensuu
Joensuu

1 Introduction

Image and virtual environment formats are more powerful and efficient tools for communication than any of the verbal or numerical formats that have preceded them. They appear to be especially useful for communicating complex data sets that involve subtle and ambiguous relationships among component parts. Forestry is an area where there has been a dramatic increase both in the complexity of what is known about the interactions within forest ecosystems, and in the necessity to incorporate multiple values in the design and implementation of management strategies. For these reasons data visualization and virtual reality technology are receiving increasing attention in forestry decision making.

Increasing use of visual interpretation of data sets and processes has been a major trend in all areas of science, with the aim of bringing greater understanding to complex problems (Cox, 1990). Researchers have widely applied data visualization to the problem of managing environmental resources (Larson *et al.*, 1988; Onstad, 1988; Orland, 1993), and in the quest to better understand human behaviour vis-a-vis those resources (Malm *et al.*, 1981; Orland *et al.*, 1992a). Visualization has also been used to depict various characteristics and variations existing in forest settings (Kellomäki and Pukkala, 1989; Orland *et al.*, 1994) and to judge the visual quality of the landscape (Thuresson *et al.*, 1996; Daniel and Meitner, 1997).

Despite increasing evidence that visual communication is enabling greater numbers of people to benefit from the power of computers, there has been little to support or caution the use of particular modes of graphic representation

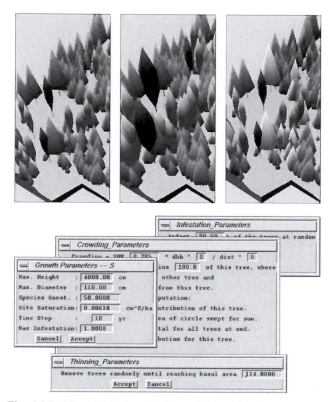

Fig. 14.1. Visual interface to forest information databases.

in different settings. Unfortunately, understanding of the role of images in human communication may be limited by the long-standing bias for verbal communication and mathematical analysis in Western society. Dynamic images and virtual environment formats are readily accepted as useful methods for effective communication of complex data. Meanwhile, there are many questions regarding how people discern meaning from these new media and whether it is necessary to acquire some level of visual literacy to be able to use such human-technology interfaces. Disturbingly, those charged with discerning the meaning of visualized results have frequently had little education or practice in the creation and interpretation of imagery. There has been some discussion of the necessity for systematic investigation of the tools but very few examples. The investigation of a virtual world user's ability to accurately extract meaning from displayed data is still largely unexplored.

1.1 Evolving Uses of Visualization in Forestry

The earlier uses of computer visualization in forest planning fell into two broad types: First, following the examples of computer-aided-drafting and Geographic

Information Systems (GIS), there have been programs using simplified computer graphic representations to illustrate the contents of an underlying database of plant species, size classes, etc. found in forest inventory databases (*e.g.* Kellomäki and Pukkala, 1989; McGaughey, 1997). Typically the emphasis of these programs has been on numerically accurate but visually symbolic representations of the database contents - they have generally not achieved high levels of visual realism (Figure 14.1).

Second, and especially for study of forest visual quality, *calibrated* photographic images have been used (*e.g.* Bishop and Flaherty, 1990; Orland *et al.*, 1994). Images are calibrated to underlying forest conditions by careful image editing guided by extensive libraries of images of ground conditions. The emphasis here has been on establishing valid representations of the visual conditions with less ability to demonstrate strong relationships to underlying tree data (Figure 14.2 and Plate 9). These two categories have been called geometric modelling and video imaging, respectively, by previous authors (*e.g.* Orland, 1993; McGaughey 1997).

Visualizations developed in either of these ways have been used widely in making presentations about projected future conditions of forests. Typically they have been incorporated into decision-making processes in the place of (or in addition to) text and numeric data. Images have thus become increasingly a part of public presentations, environmental impact statements, forest recreation surveys and, lately, online computer information systems. However, in general, the mode of presentation has been one of delivery only. Data generated and owned by a forest management agency is processed to create a set of visual materials illustrating the issue in question. The visualizations are presented to represent a forest management issue, discussions ensue, and a decision is made to address or resolve the issue. In each case the role of the visualization, while important, is intended to illustrate issues in an impartial and detached manner.

In recent years there has been increased emphasis on ecological mathematical modelling as a major activity within the scientific community, and on the representation of the modelled results via visualization. Tools such as Geographic Information Systems, for storing and manipulating spatially-referenced databases, include visualization tools that combine some of the attributes of the earlier tools. With a visual interface capable of communicating complex ecosystem interactions to users, the potential arises for going beyond interfaces that simply display the contents of existing forest information databases. For instance, users might be able to predict for themselves the effects on a wildlife population of manipulating forest stand density, average diameter at breast height (dbh; *i.e.* 1.3 metres above the ground)) and canopy closure by applying a dynamic ecosystem model (Orland *et al.*, 1997). Such models enable users to interact with simulations of system dynamics and conduct "thought experiments" as they explore different resource changes (Figure 14.3).

Accompanying the potential for interaction and exploration of data relationships, the graphic performance capabilities of personal computers have dramatically improved, enabling faster, more versatile and realistic visualizations

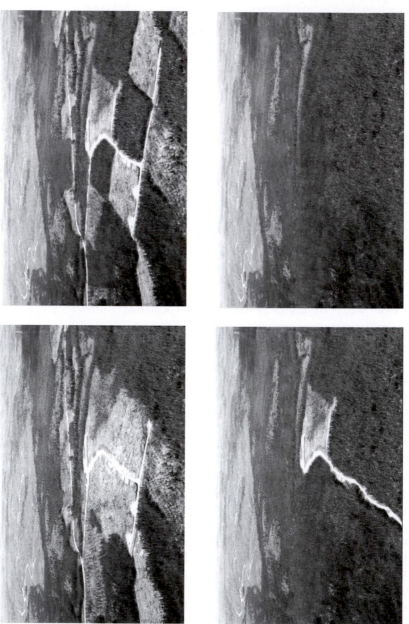

Fig. 14.2. Visualization of proposed forest harvest patterns in Northern Ontario, Canada (see Plate 9).

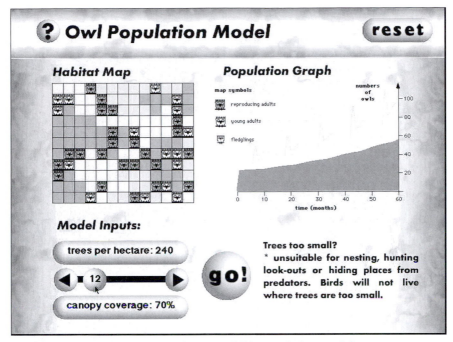

Fig. 14.3. Game-like visual interface to wildlife population model.

to be developed than had previously been possible. Highly realistic images can be created either by texture-mapping techniques or by sophisticated three-dimensional growth algorithms. Interfaces have moved from simple two-dimensional displays, through three-dimensional representations that can be rotated at will, to fully immersive worlds through which the user can navigate at will. The most intractable technological restrictions to the construction of immersive virtual forests, integrated with existing forest management and spatial databases, are the size and speed of current computers. At the scale of complete landscape vistas, forest databases are very large and very complex. Simple mathematics indicates that even quite modest vistas encompass many millions of trees, each of these with scores of attributes.

There are numerous areas where virtual reality (VR) technology has been predicted to revolutionize traditional teaching, planning or management tasks. This paper questions, however, whether in the move from static two-dimensional graphic displays to fully immersive VR there are significant differences in the way we address and evaluate the displayed information - an interaction between display mode and the effectiveness or validity of the display in supporting management decisions. We will discuss and categorize different features of visual interface design and of forest management, and then analyse to what extent the attributes of VR resonate with the demands of modern forestry. As a case study we will briefly describe and evaluate the attributes of a "first generation" virtual environment

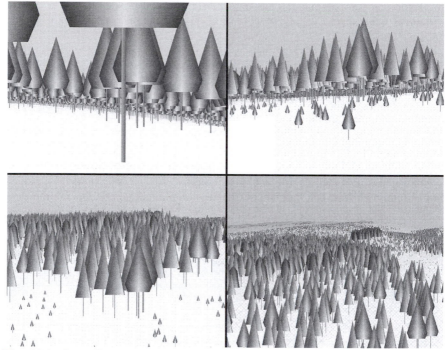

Fig. 14.4. SmartForest virtual forest interface.

- the SmartForest forest visualization tool (Orland, 1994; Uusitalo *et al.*, 1997) being developed at the Imaging Systems Laboratory, University of Illinois (Figure 14.4).

2   General Characteristics of an Information Interface

In 1992 Orland wrote of a visualization dilemma. While strategic planning of management activities occurred at large scale and regularly employed the use of maps and overview kinds of material, active management occurred at small scales where overview maps would be quite unsuitable. Such management instead demanded ground-level, three-dimensional, and small scale visualization support. Other authors have termed this distinction Exocentric-Egocentric, the distinction referring to the viewpoint of the observer either from outside or from within the apparent world of the visualization (Howard, 1991). It is proposed here that the significant difference is in the degree of immersion achieved in the interface and that there is an important correlation between the active engagement with the environment achieved by the user of the interface and the experience of apparent immersion of the observer in the simulated environment. Further, it is proposed that this ability to experience engagement and immersion introduces expectations for environmental communication that may not have existed previously, and that may or may not be addressed by the new visualization technologies. In other words, introduction of the ability to interact with a virtual world changes the expectations and responses users will make in subsequent decision-making situations.

The effectiveness of any user interface will be based on its ability to achieve a range of communication goals. While environmental agencies have previously focused on visualization for internal communication, that focus has shifted to include the general public for whom satisfaction with the visualization interface may be critical. If the public are not satisfied with their interaction then the information interface may not be serving the public review process. Thus far there are few published critical evaluations of models of information delivery via computer-mediated interfaces. Morris and Ogan (1996) have proposed some criteria for establishing successful information delivery based on basic principles of communication theory:

- *Interactivity*
  The more effective the interaction between user and information delivery system, the more likely users are to engage in the exchange, and the more successful will be the communication. In applying the concept to a virtual environment interface, opportunities for interaction would thus be viewed as valuable, since they achieve more fully interactive communication.

- *User Gratification*
  More effective communication takes place when the user's information desires are satisfied by the encounter. This may include direct access to original data, means of understanding basic ecological concepts, and ways to ask questions of responsive experts.

- *Social Presence and Media Richness*
  While it is critical to avoid trivialization of the issues being portrayed, or the unintentional demeaning of the user, it may be helpful to imbue the interface with a personality. Social presence theory attempts to determine the different abilities of communication media to engage the user through the degree of social cues inherent in the technology. In general, computer interfaces with their relative lack of direct visual and other non-verbal social cues, are low in social presence in comparison to face-to-face communication.

These widely accepted *marketplace* dictates for successful information interface design have implications for the design of forest visualization interfaces in as much as they point towards an increasing intent to engage the information user with the information content.

3   Interface Issues in Developing a Virtual Forest

Virtual reality applications combine fast computer graphics systems with display and interface devices that provide the effect of immersion in an interactive three-dimensional environment in which the objects have spatial presence (Bryson, 1995). In 1997 the National Research Council report, *Modeling and Simulation: Linking Entertainment and Defense* (Zyda and Sheehan, 1997) described opportunities emerging from interdisciplinary development in visualization and virtual reality that might revolutionize the ways in which scientists and the public approach the comprehension of complex scientific data. Two areas were highlighted as having special potential - virtual presence and persistent virtual worlds - the expressions referring to the abilities of VR environments to immerse users in environments so they feel as if they are "there" and to thereby evoke real-world responses and performance. Such three-dimensional computer generated environments are generally called virtual environments (VE). Presenting data in a three-dimensional format does not in itself define VR. Although there are no generally approved definitions for VR there are several criteria that are commonly linked with the term VR (Wann and Mon-Williams, 1996), as follows:

- *Perception of Three-dimensionality*
  The virtual environment should provide realistic depth information that enables observers to judge the correct three-dimensional relationships of virtual objects.

- *Perception of Immersion*
  The virtual environment should provide the illusion of immersion within the environment. The interface should enable the user to navigate within the environment in all dimensions and traverse the world as if it were real.

- *Interactivity*
  The user should be able to interact with the displayed data through simple actions rather than at the level of program and command structures. The interface should minimize the distance between the thought processes of the user and the operation of the system (Hutchins *et al.*, 1986).

- *Media Presence*
  Motion and interactions should provide the illusion of being in a real environment - virtual presence - wherein the environment responds as the user acts, without temporal or spatial delays (Bryson, 1995).

While the requirements for an information interface described above are necessarily very general, it can be seen that they map quite well to the proposed criteria for a virtual environment. Thus a well-designed virtual forest interface might be expected to address basic communications needs and provide the user a satisfying encounter with forest information.

4 Characteristics of Forest Management and their Compatibility with Virtual Technology

The discussion to this point has focused on the general applicability of principles of communication to forest visualization. There are also characteristics of forest management that are not only addressable by visualization tools, but for which such tools are uniquely powerful ways of investigating otherwise intransigent problems in communicating about forest change.

4.1 Time-Scale

The time-scale of forest growth and development is vastly different from the time-scale of human management actions. Actions may only occur once in the lifetime of a forest manager but must still be accomplished at the appropriate time and with concern for the future implications of the changes. The timing of those changes may have significance for a wide range of proposed forest developments. In all these instances visualization can greatly aid the evaluation of alternative long-term plans. Visualization enables the observer to perceive changes in the forest, and to communicate the extent and severity of major environmental changes, such as insect outbreaks, without time limitations. Future forest growth can be simulated to show how the forest will look after any user-defined time period. These projections facilitate discussions of treatment alternatives and promote better understanding of natural and man-made stand conditions. In these kinds of situations, visualization can be vital in developing awareness of emerging problems, and in motivating agencies and individuals to respond to these problems (Orland, 1994). There are currently few other methods than to use a combination of visual and time-based media to attempt to communicate effects occurring over long periods of time, or anticipated in the distant future.

4.2 Irreversible Decisions

Within the life span of human beings, many processes of environmental change are essentially irreversible. Forest harvesting is especially visible in the landscape and may have huge impact on scenic resources and recreational activities as well as on the future growth and yield of the timber resource. Although forestry is generally guided or constrained by governmental legislation or recommendations, the decision-maker always has several alternative strategies to choose from in terms of logging methods and intensities. The use of a virtual forest provides the possibility of assessing the consequences of each alternative before they ever occur. The opportunity to pre-view and trade-off alternatives is critical in areas with high scenic beauty, economic or ecological value.

4.3 Spatial-Quantitative Variation

Information about the contents of each forest stand has often been presented as simple mean values. However, forests also exhibit between-stand-variation of important tree characteristics relevant to managing the forested landscape. The forest landscape may also possess rich diversity of wood lots, roads, agriculture land, water elements, buildings, etc. which all have to be taken into consideration while making management decisions. Timber managers are under pressure to seek out desirable species mixes and dimension and quality characteristics within the forest while maintenance of biological diversity requires that different forest structures be provided within each forest holding under active management.

The major benefit of displaying large-scale forest data via a virtual world interface is that of being able to simultaneously display numerous channels of data and hence be able to perceive and interpret interrelationships between those data items. The location of data items in space is clearly of vital interest to ecological and other scientists. The kind of relationship which is best suited to visual display within a multi-dimensional virtual space is also a critical issue.

4.4 Multiple Objectives

As a result of the growing interest in non-economic forest attributes, new research has been focused on the incorporation of models of scenic beauty (*e.g.* Brown and Daniel, 1986), recreational amenity, and wildlife habitat (Kangas *et al.*, 1993), as well as various biodiversity indices. A virtual forest explicitly displays multiple channels of data simultaneously. Visualization is perhaps the only effective means of presenting such multi-variate data and the virtual forest enables the decision-maker to rapidly synthesize this complex information in an easily understood format.

Even deeper engagement with the data might be achieved by interacting with the visual display. The user should have the possibility to query the data according to any user-defined classification. It should be possible to query each object within the virtual forest and change its value. Databases can be linked with any advanced forest management tool providing users with the possibility to contrast consequences of proposed activities with monetary benefits or changes in wildlife population.

5 Data Quality Issues in Developing a Virtual Forest

To what degree are the data commonly available within forest information systems able to support the development of virtual environment interfaces? A virtual forest interface is both a consumer of forest spatial information and a delivery vehicle. The interface is explicitly spatial and requires that all data be presented in a spatial context. This condition both provides opportunities and reveals constraints in the potential for management use of a virtual forest.

Constructing a virtual forest requires integrating information from various sources, information about the topography, delineations of management boudaries such as stands, and forest inventory data possibly including the outputs of forest growth simulators and other ecosystem models.

In most forest visualization applications, the terrain elevation is in the form of a digital elevation model (DEM). The format of the US Geological Survey is typical. The USGS DEM presents an array of spot elevations at 30-metre spacings on a north-south/east-west regular grid. There are significant issues of registration that derive from fitting a regular rectangular grid to a curved surface and techniques for handling the situation have been developed. Of more significance to this chapter is that the grid be viewed as a sample of the terrain, not a complete representation, if such a thing could exist. Elevations falling between sample points must be extrapolated from those points and thus will rarely match true ground conditions. The use of a 30-metre grid resolution is a response to the burden of collecting and handling data for the vast areas of any geographic domain. Although a convenience for data collection, at the ground level and in the viewer-terrain relationship expected in a virtual environment, the 30-metre grid size is much greater than human scale. Relying on simple interpolation of elevations can result in unconvincing quarter acre (0.1 ha) slabs of gently sloping terrain.

The nature and spatial distribution of forest cover is delineated into relatively homogeneous areas of forest cover called stands. Most applications read information about such homogeneous stand boundaries in some grid format. Again, the issue of relevance here is the conflict between the need to rationalize the collection and manipulation of inventory data vs. the need of a virtual environment for spatially explicit location of objects. A natural system is not homogeneous. Rather, the assessment of homogeneity is a judgement made to bring economies to data collection and to enable projections to be made in an expedient way. Homogeneity in this case is scale-specific. At the level of the forest landscape it is economical to describe stands as homogeneous units. At the level of the proposed visitor to a virtual forest environment, the same stand is far from homogeneous.

Topographic information from the DEM, and the vector boundaries of the stand delineation, provide the basic spatial information for visualizing forest data. Assigning objects to spatial coordinates is a critical next step in developing visualizations. The earliest virtual forest visualizations treated data items (i.e. trees, stones, houses, etc.) quite literally. SmartForest-I (Orland *et al.*, 1994) used stem-by-stem forest stand inventories as their data source. Each individual tree was located by the xy coordinates of its base and was scaled by the stored values for height, trunk diameter and crown shape. However, such data sets for real forests are so extensive and item-by-item inventory so expensive that developing such visualizations would be prohibitively expensive. More usually data will be collected by various sampling methods, summarized, and later represented by techniques involving the probabilistic reconstitution, expansion, and spatialization of the data. Trees in a stand are typically described by a list of 10-150 surrogate trees - each surrogate standing in for the whole of a class of *n* close-to-identical trees. A complete list of each tree in a stand is rarely available, and in most cases unnecessary, since trees are generally not managed with a high degree of specificity. In contrast, homes, power lines and road surfaces do demand precise representation of direction and cartographic location - data that is normally not part of the sample-based database of forest components. Incorporation of such elements requires more sophisticated means for assigning spatial data.

The virtual environment point-of-view presents information as if real. No differentiation is made between objects that are known to be of a particular shape and location vs. others that are assigned shape and location as an estimate based on a data sample. The concept of immersion and engagement with the environment becomes problematic when the environment is built of objects determined by statistical rather than concrete spatial relationships.

6   Case: SmartForest

At this point it is evident that a virtual forest interface may offer considerable advantages as a means of communicating forest resource information. It is also evident that virtual forests can be developed that satisfy the defining characteristics for immersive environments. While the spatially specific data needs of a fully accurate immersive environment cannot be met, there are numerous efforts underway to provide data of sufficient quality to support experiences that represent the qualities of the projected environment vs. the actuality of that real-world environment. Finally, a number of areas have been suggested in which there is no apparent alternative to the use of a complex visualization environment as a decision-making medium.

However, while visualization clearly offers benefits to the forest scientist and manager, issues arise in the development of visualization systems that place limitations on the usefulness of the tools - or provide unanticipated bonuses in solving other problems. We provide an analysis of one typical visualization tool, SmartForest, in order to illustrate these points. SmartForest is one of a growing class of forest visualization tools, some directed at displaying visually realistic forests, others focusing more on scientific visualization (Figure 14.5 and Plate 11).

SmartForest is an advanced graphically oriented forest visualization program developed at the Imaging Systems Laboratory, University of Illinois in collaboration

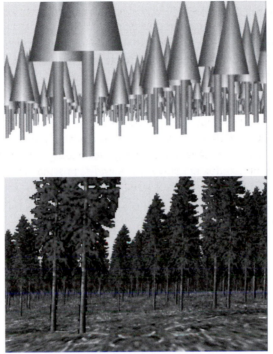

Fig. 14.5. SmartForest views. Above: iconographic (management mode). Below: realistically textured (landscape mode) (see Plate 11).

with the USDA Forest Service and the University of Helsinki. SmartForest possesses advanced tools for moving and interacting within and with a forest setting. The user can view the ground level within a forest, walk between the trees, view large forest areas from user-defined aerial height and classify stands and trees by highlighting them with different user-defined colour-codes. With separate commands the program renders the ground and trees with bit mapped textures sampled from forest photographs that create a more realistic virtual world surrogate of the forest. SmartForest was originally written in C, using the OpenGL library of graphic primitives and X-windows/Motif (Orland, 1994; Uusitalo *et al.*, 1997) but has also been ported to the PC/Windows platform (Uusitalo and Kivinen, 1998).

6.1 Presentation of Forest Information

SmartForest operates in two different modes: management mode and landscape mode. Management mode is a simplified presentation of the real forest conditions that helps the manager quickly and efficiently query and analyse the different characteristics of the forest stands and single trees. Landscape mode is a realistic presentation of the real world. Trees are presented as texture-mapped objects and the ground is wrapped with realistic two-dimensional ground images. Landscape mode facilitates the evaluation of the visual effects of different forest harvest practices.

6.2 Valid Depiction of Real World Conditions

SmartForest renders a depiction of the real forest that corresponds closely with underlying data structures. A digital elevation model (DEM) provides topographical data for creating the landform features; a stand file provides the locations for stand data to be overlaid on the DEM; and tree list files provide the records of surrogate trees to place in those stands. Each surrogate tree is given a weight that indicates the number of those trees in a stand. The program distributes trees within the grid cells defined by the stand files, using a random number generator to assign trees to locations within the grid, controlled to ensure a minimum 1.0-metre between-tree distance. The program can also represent regular grid or row plantation planting.

The program can be used to visually classify trees according to many different tree characteristics. Data selection tools can be used to customize a classification and define colour palettes to represent specific resource qualities. For procurement officers and mill-owners these features may be critical in determining the feasibility of particular purchasing or harvesting plans. The specific information for a single

Fig. 14.6. Tree canopies shaded by trunk diameter (diameter at breast height) (see Plate 10).

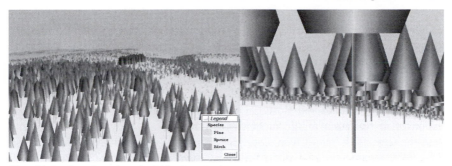

Fig. 14.7. Overhead and within-forest views.

tree can at any time be queried by a simple mouse click, resulting in a pop-up menu of general characteristics of the tree (i.e. species, dbh, tree height, number of trees in the stand, etc) (Figure 14.6 and Plate 10).

SmartForest uses data sources typically used for day-to-day forest planning. To reconstruct data from a sample-based inventory, such as those described above, demands that spatial characteristics be synthesized. Commonly this is achieved by random assignment of trees to points on a regular grid. Inevitably there is a mismatch between the apparent veracity of the displayed image and the acknowledged lack of spatial specificity in the data represented. The ability of the virtual world user to engage the data if anything exacerbates this issue - while the apparent realism of the data objects to the user is enhanced, the likelihood of that data accurately representing specific on-ground conditions is remote.

6.3 Freedom of Movement, Sense of Presence

SmartForest has been developed to enable a variety of ways to move in a forest setting. The user can manually input a *x, y, z* spatial location and a direction of view. The target area may be located in a map window and then visualized by clicking the desired viewing point. Finally, for even greater interaction it is possible to move by dragging the cursor within the view while holding different mouse buttons to achieve longitudinal, rotational, and vertical movement. Provided the program is run on a sufficiently powerful workstation or PC, an immersion effect - that the user really walks in the forest - can be achieved. However, at this time the forest within which the user is immersed is an abstract and symbolic representation of forest conditions. While the landscape mode of display achieves some realism it is only useful as a final presentation tool - current computers are not fast enough to provide that realism as the visitor walks through the virtual forest (Figure 14.7).

As computing capabilities improve, the graphical tree-icons of the present interface will gradually be replaced by more realistic representations. At this time, though, there is no illusion in the mind of the visitor that they are in a real forest. Presenting the knowledge that this is sampled data is fitting to the diagrammatic quality of the interface - there is no suspension of belief to the extent that the visitor would believe in the reality of the objects in the scene. The danger in increasing the realism of the scene is that evaluation of its conditions will move from a conscious, cognitive, weighing of the issues represented by the symbolic display of data, to a more emotional, affective, response to the scene as if it were real. Ulrich (1983) has proposed that the initial affective response to a scene predisposes an observer in the subsequent cognitive choices they will make. Although Ulrich and others (*e.g.* Daniel and Boster, 1976) would support the value of affective responses in reflecting public judgements, nevertheless, they may be different from the conscious, data-driven, cognitive responses intended as the outcome of using a VR interface as a surrogate for the real world.

7   Discussion

Visualization interfaces for forestry typically tackle two complex tasks - systematically representing a range of issues of management practices, and communicating those issues to an audience of non-specialists. Of the range of visualization tools available, the critical differences and advantages will lie in the interactivity and deliberate design to make the user better able to understand the relationship between underlying data relationships and forest planning at a range of scales - from the ground-level, egocentric, view to the synoptic, exocentric, overview. New tools developed recently fall within the definitions of VR devices. As shown in this chapter, VR offers promising new possibilities in aiding the evaluation of forest resources as well as alternative management plans.

With the recent acceleration of graphic performance capabilities of personal computers, there are reasons to expect that 'virtual' forest management will increasingly be expected to replace existing forest management procedures. Unfortunately, the biggest obstacles to applying visualization in practice, the lack of appropriate information as well as the difficulty of combining information from different sources and formats, will remain. Improvements in data collection will be slow and local forest management plans will continue to be based on mean values of different tree characteristics, whereas full utilization of virtual forest interfaces would require reliable information on diameter, height and quality distribution of each species.

However, the authors propose that there is a pressing need beyond the need for better data that should be addressed in the implementation of virtual forest management: that is, that virtual reality tools change more than simply the physical interface through which we view forest data. The contrast between the new tools and their predecessors demands close scrutiny because they do not represent a simple evolutionary step but more a paradigm shift in the way we address forest management information.

The GIS-based and digital imagery tools described at the beginning of this chapter share an important characteristic - they are presentation media. That is, they result in maps and images that are intended to be scrutinized from a distance in a detached professional evaluation setting. Partly because of the relatively primitive technologies they employ, there was never any expectation on the part of the user to become closely engaged in the depicted environment - and thus the qualitative value of being in the depicted place had to be guessed at rather than directly experienced. This detachment has been quite appropriate to a decision-making arena in which the goal has been to achieve rational, and detached, decision outcomes.

The general qualities proposed for a successful information interface, in contrast, subscribe to a view of communication as a much more intimate transaction between information source and information user. The emphases on interactivity and social presence of the medium are directed at the necessity to understand the information at a deeper level than by simply "reading" a map or chart. The notions of gratification and social presence point toward the development of interfaces that more actively advocate for the information they present than merely providing information. Advocacy, however, remains outside the mission of the vast majority of potential users of virtual forest interfaces. For those users the goal must be to provide, as far as is practical, impartial information, not advocacy.

Based on the guidelines for effective communication outlined above, one would expect the criteria for a successful virtual environment interface to be directed towards achieving social presence and close engagement of user with the interface, presently with little explicit attention to the validity of the resulting imagery. In the most common setting represented as virtual environments - interactive games - the goal of the interface is to induce the visitor to immerse themselves and become fully engaged in the illusion of the virtual environment. That is, the environment developer wishes the visitor to believe they are actually within the new environment and to behave accordingly. The success of tools such as flight simulators, and lately medical procedure simulators, indicates that this state is achieved quite readily. For forest managers, knowledgeable about the quality of their data inventories and projections, the goal should be to heighten healthy scepticism rather than to suspend belief.

It is all too common to read the call from university scientists for developing a research agenda, and even more common to hear that "more research is needed." In this case it seems that the guiding principles for developing virtual environment interfaces that are emerging from communications science are very clear. However, in part the goals for those design principles may run counter to the needs of agency forest managers who need to tell a detached and impartial story. The design criteria for virtual reality environments are equally clear, but again may run counter to the need to clearly represent the uncertainty, ambiguity, and indeterminacy of data and projections of future forest conditions. If we are going to gain the benefits the emerging technologies appear to offer for forest decision-making, we had better be ready to identify where we need to proceed with caution and develop mechanisms to ensure the appeal of the tool does not subvert our goals of rational, open and responsive management.

Acknowledgements

We acknowledge the following contributors to the development of SmartForest. Ross Pywell, Jeanine Paschke of the US Forest Service FHTET; Veli-Pekka Kivinen of the University of Helsinki; Abhijeet Chavan, Paul Radja, Kittipong Mungnirun, Kenneth Schalk, Kaiyu Pan and Kun Liu of the Imaging Systems Laboratory, University of Illinois.

References

Bergen, R.D., C.A. Ulricht, J.L. Fridley and M.A. Ganter (1995) The validity of computer-generated graphic images of forest landscape. *Journal of Environmental Psychology* 15(2): 135-146.

Bishop, I.D. and E. Flaherty (1990) Using video imagery as texture maps for model driven visual simulation. In: *Proceedings Resource Technology 90*, Washington, D C. ASPRS, Bethesda, Maryland, pp. 58-67.

Brown, T. and T.C. Daniel (1986) Predicting scenic beauty of timber stands. *Forest Science* 32(2): 471-487.

Bryson, S. (1995) Approaches to the successful design and implementation of VR applications. In: R.A. Earnshaw, J.A. Vince and H. Jones (eds) *Virtual Reality Applications*. Academic Press Inc., San Diego, pp. 3-15.

Cox, D.J. (1990) The art of scientific visualization. *Academic Computing* 46: 20-56.

Daniel, T.C. and R.S. Boster (1976) *Measuring Landscape Esthetics: The Scenic Beauty Estimation Method*, USDA Forest Service Research Paper RM-107, Rocky Mountain Forest and Range Experiment Station, Ft Collins, CO.

Daniel, T.C. and M.J. Meitner (1997) Predicting human response to future environments through data visualizations. In: *Proceedings 1997 ACSM/ASPRS. Seattle, Washington*. American Society for Photogrammetry and Remote Sensing, Bethesda, Maryland. Technical papers volume 4, 276-287.

Howard, I.P. (1991) Spatial vision within egocentric and exocentric frames of reference. In: S.R. Ellis (ed), *Pictorial Communication in Virtual and Real Environments*. Taylor and Francis, London.

Hutchins, E.L., J.D. Hollan, and D.A. Norman (1986) Direct manipulation interfaces. In: D.A. Norman and S.W. Draper (eds) *User Centered System Design*. Laurence Erlbaum, Hilsdale, New Jersey, 87-124.

Kangas, J., J. Karsikko, L. Laasonen and T. Pukkala (1993) A method for estimating the suitability function of wildlife habitat for forest planning on the basis of expertise. *Silva Fennica* 27(4): 259-268.

Kellomäki, S. and T. Pukkala (1989) Forest landscape: A method of amenity evaluation based on computer simulation. *Landscape and Urban Planning* 18: 117-125.

Larson, S.M., G.R. Cass, K.J. Hussey and F. Luce (1988) Verification of image processing based visibility models. *Environmental Science and Technology* 22(6): 629-637.

Malm, W., K. Kelley, J. Molenar and T. Daniel (1981) Human perception of visual air quality (uniform haze). *Atmospheric Environment* 15(10/11): 1875-1890.

McGaughey, R.J. (1997) Visualizing forest stand dynamics using the stand visualization system. In: *Proceedings ACSM/ASPRS/RT, Seattle, Washington*. American Society for Photogrammetry and Remote Sensing, Bethesda, Maryland, Technical papers volume 4, 248-257.

Morris, M. and C. Ogan (1996) The internet as mass medium. *Journal of Computer-Mediated Communication*, Vol. 1, No. 4. Online journal: <www.ascusc.org/jcmc/vol1/issue4/morris.html>.

Onstad, D.W. (1988) Population dynamics theory: The roles of analytical, simulation, and supercomputer models. *Ecological Modelling* 43: 111-124.

Orland, B. (1992) Evaluating regional changes on the basis of local expectations: A visualization dilemma. *Landscape and Urban Planning* 21: 257-259.

Orland, B. (1993) Synthetic landscapes: A review of video-imaging applications in environmental perception research, planning, and design. In: D. Stokols and R. Marans (eds) *Environmental Simulation: Research and Policy Issues*. pp.213-252.

Orland, B. (1994) SmartForest: 3-D interactive forest visualization and analysis. In: *Proceedings Decision Support-2001, Resource Technology 94*, Toronto. American Society for Photogrammetry and Remote Sensing, Bethesda, Maryland, pp. 181-190.

Orland, B., J. Vining and A. Ebreo (1992a) The effect of street trees on perceived values of residential property. *Environment and Behavior* 24(3): 298-325.

Orland, B., T.C. Daniel, A.M. Lynch and E.H. Holsten (1992b) Data-driven visual simulation of alternative futures for forested landscapes. In: *Proceedings IUFRO - Integrating Forest Information over Space and Time*. Canberra, pp. 368-378.

Orland, B., T.C. Daniel and W. Haider (1994) Calibrated images: Landscape visualizations to meet rigorous experimental design specification. In: *Proceedings, Decision Support 2001 - Resource Technology 94*, American Society for Photogrammetry and Remote Sensing, Washington, DC. pp. 919-926.

Orland, B.C. Ogleby, I. Bishop, H. Campbell and P. Yates (1997) Multi-media approaches to visualization of ecosystem dynamics. In: *Proceedings 1997 ACSM/ASPRS*. Seattle, Washington. American Society for Photogrammetry and Remote Sensing, Bethesda, Maryland. Technical papers volume 4. pp.224-235.

Thuresson, T., B. Näsholm, S. Holm and O. Hagner (1996) Using digital image projections to visualize forest landscape changes due to management activities and forest growth. *Environmental Management* 20(1): 35-40.

Ulrich, R.S. (1983) Aesthetic and affective response to natural environment. In: I. Altman and J. Wohlwill (eds) *Behaviour and the Natural Environment*.

Plenum, New York, pp. 85-125.

Uusitalo, J. and V-P. Kivinen (1998) Implementing SmartForest forest visualization tool on PC environment. In: *Proceedings Resource Technology '98 Nordic, Rovaniemi*, Finland. The Finnish Forest Research Institute Research Papers <http://www.metla.fi/event/rt98/abs/Uus-Kiv.htm>.

Uusitalo, J., B. Orland and K. Liu (1997) A forest visualization interface for harvest planning. In*: Proceedings ACSM/ASPRS/RT Annual Convention*. Seattle, April 7-10, 1997. pp. 204-215.

Wann, J. and M. Mon-Williams (1996) What does virtual reality NEED?: human factors issues in the design of three-dimensional computer environments. *International Journal of Human-Computer Studies* 44: 829-847.

Zyda, M. and J. Sheehan (1997) *Modeling and Simulation: Linking Entertainment and Defense.* Committee on Modelling and Simulation, National Research Council, National Academy Press, Washington, DC.

Chapter fifteen:

Considerations for Digitial Visualization of Landscape

John Danahy
Centre for Landscape Research and the School of Architecture and Landscape Architecture, University of Toronto
Toronto, Ontario

1 Introduction

This chapter is concerned with the issue of selecting and using visualization techniques for visual resource management (VRM) and environmental decision-making. The capacity of visualization technology to systematically represent both complex landscape scenes and abstract phenomena has progressed to the point where issues of technological capability are no longer as fundamental as they were a decade ago. Now that there is a diversity of high quality visualization options available, the true power of digital representations lies in properly matching them to our capacity to systematically associate abstract knowledge representations with visually explicit representations. Particular attention should be paid to matching the purpose of the visualization model to the structure of its data and the graphic rendering techniques of computer systems. In contemporary issues of forest management, aesthetic and ecological models could be related via the visualization. The spatial model in a visualization can provide a systematic link between symbolic and sensory understandings of the environment.

This chapter also provides an overview of the progression and synthesis of the two most commonly applied types of computer graphics found in landscape visualization (image processing and synthetic geometric representation). It identifies thresholds and breakthroughs afforded by visualization technology that make computed representations of landscape both credible and timely enough for use in visual resource management (VRM) and environmental decision-making. The chapter predicts that the continued convergence of image processing and synthetic geometric representation combined in highly interactive systems for decision-making will gradually alter practice and heuristics regarding the credibility of images (see Chapter 13 in this volume for further discussion). The next generation

225

Fig. 15.1. Synthetic image of unbuilt resort proposal by Frank Lloyd Wright for Emerald Bay, Lake Tahoe. Credits: CLR, U of T, Canadian Centre for Architecture. Synthesized using CLR's PolyTRIM software (see Plate 12).

of interactive knowledge media will improve our ability to enter into two-way communication between experts, decision-makers and the public.

The final section of the chapter presents a matrix of abstraction considerations and heuristics used at the Centre for Landscape Research (CLR) to guide the production of interactive visualization projects. This matrix identifies and relates a series of parameters that influence the effectiveness and cost of interactive real-time visualizations. These parameters are illustrated using examples from VRM and environmental decision-making projects undertaken at the CLR through the 1980's and 1990's (such as the model illustrated in Figure 15.1). The figures in the chapter are representative of what one can expect to create on personal computers in the coming years.

2 Visual Literacy Prosthesis

There is an essential need for visualization simulations to be fair and credible in the minds of the people affected by a VRM decision. If an image is overly abstract it may leave too much room for misinterpretation (Sheppard, 1989). In practice, most VRM studies are conceptually straightforward exercises. The process is designed to ask for peoples' approval on a proposed course of action or choose between professionally derived alternative plans. In this context, it is desirable to minimize the potential for misinterpretation. As such, the assumption used in practice is that the more explicit the representation is made the better the idea will be communicated. This is only true if the idea is well developed. An ambiguous conceptual idea communicated with an image that is too explicit is likely to be misleading. The explicit didactic nature of a photorealistic image can limit not only the potential for misinterpretation but also the potential for creative imagination. In a creative planning exercise it is common practice to use the ambiguity of an abstract representation to avoid a premature limitation on the scope of ideas considered during conceptual stages of problem solving.

Conventional VRM practice assumes that it is too time consuming, expensive and potentially too confusing to have laypersons involved in the search for fundamental questions and solutions. Instead, citizens are usually asked to choose between options and are removed from the technical aspects of searching for solutions. On occasion, citizens are sampled for preferences at the outset of a project to provide experts with a general profile of the users. The assumption behind this approach is not unwise, given the lack of visual literacy found in the general population when compared to that of trained professionals. The average person may be quite capable of reading an image but completely unskilled at expressing their own ideas and issues directly in a visual medium. In contrast, the gap in verbal and written language skills between professionals and laypersons is relatively small. Citizens regularly read written reports and prepare written comments or verbal presentations for submission to decision-making processes. Visualization technology manifests a potential that may change this situation.

As with conventional spoken and written languages, visual and spatial literacy skills take years to develop. Digital visualization technologies offer some prosthetic potential in shortening the visual literacy gap between laypersons and experts. It is suggested here that when people see ideas in a more explicit form, closer to a direct replication of the sensory stimulus one would experience in the real landscape then it should be easier for people to understand the visual implications of a planning or management strategy in the landscape.

Visualization gives people a chance to think about or read visual material in thoughtful ways similar to the way one can use language to contemplate the ideas in a text. This is particularly true of interactive visualizations where the viewer can ask to change a parameter such as viewpoint or lighting condition in the scene. This type of request is virtually impossible to fulfil when images require significant amounts of time and expertise to effect. With the ability to learn about the characteristics of a scene by changing the syntax of its elements (vocabulary) people become more literate with the image as a form of communication. People can more consciously think about the issues and the solutions. If used in this way, the technology can act as representational prosthesis to help a person rapidly give form to thoughts.

3 Usability of Computer Graphics to Represent Landscape

The last twenty years have seen computer graphics progress to a level where synthetically computed visualizations of landscape convincingly approximate much of what a camera can record in a real landscape. In general practice, people seem prepared to accept the synthetic photorealistic images produced by specialized visualization systems developed during the late 1990s. The images from these technologies are being accepted as sufficiently realistic for use in large-scale visual resource planning and environmental decision-making. Jurisdictions such as the province of British Columbia are accepting synthetic images as part of the VRM process (Sheppard, 2000). A dissertation by Dr. E. Lange at the ORL Institute, of the Swiss Federal Institute of Technology demonstrated the willingness of professionals and laypersons to accept synthesized representations of landscape as equivalent to photographic depictions of the same landscape (Figure 15.2) (Lange, 1999).

The opportunity this technology presents to completely synthesize a photorealistic representation of a landscape in a systematic computable medium is highly significant to VRM practice. It means that automation of the procedures used in visualization can be applied across the full spectrum of representation requirements (from abstract to exact depiction). Automation in image production holds the promise of removing time and cost barriers to the widespread application of visualization in day-to-day management and planning (Danahy, 1999).

The basic technological barriers in visualization are now largely understood and the equipment needed to compute credible images is affordable and readily available to create single frame imagery. Commercial software such as *World Construction Set* runs on personal computers and is being actively used in forest management practice. The next stage in the evolution of visualization will be the introduction of highly interactive virtual reality visualization capability on personal computers. Real-time interactive graphics boards for personal computers

Fig. 15.2. Dr. Lange's study empirically compared real photographs with virtual images synthesized using the PolyTRIM software. The conclusion suggests that both professionals and laypersons are prepared to accept this type of visual simulation as credible for use in planning. Credits: ORL Institute, ETHZ.

developed for gaming applications now possess the texture memory architecture needed to support the applications of visualization illustrated in this chapter. The greatest barriers to the general use of visualization across a broader array of situations will involve the culture and heuristics of VRM practice.

To this point, most visual resource management studies have used two quite different types of computer graphics to visualize landscape. The introduction of personal computer based image processing techniques in the early 1980s made it possible to superimpose visual change onto digitized photographs. The approach is a digital equivalent to conventional manual photo-montage. This type of visualization tool has been used in practice when a high degree of visual realism was required (Sheppard, 1989; Orland, 1993). Image processing has made it possible to produce photographic quality visualizations with cost-effective technology (Figure 15.3).

Image processing relies on information gathered from the real world by sensors such as modern CCD (charge-coupled device) technology in camcorders or the digital scanning of chemical sensing material such as film. It requires the wise selection of representative viewpoints since the base imagery cannot be resampled in the midst of a decision-making session. The introduction of proposed or predicted change into the image needed for VRM projects requires both the skill and judgement of an expert illustrator to prepare. Most visualization projects of this type require significant amounts of operator time to complete. Great care must be taken if the change added to the image involves a change to the perspectival optical structure of the transformed image. The potential for computation to amplify an operator's skill in this task is limited. Even the more advanced image processing software available today has not sped up the process enough to see its widespread application in every day planning. The manual preparation time involved in image processing for VRM exercises would need to be significantly reduced before this could happen. It may take a reduction in preparation time of one or two orders of magnitude to match the limited time available to participants in the

Fig.15.3. Image processing photo simulation of proposed strip-harvesting at Gavin Lake in British Columbia. Credit: John Lewis and Stephen Sheppard, Collaborative for Advanced Landscape Planning, UBC, Vancouver (see Plate 14).

average planning and management process. In short, the approach is not scalable by computation.

The second type of computer graphics synthesizes an accurate perspectival image from a palette of geometric descriptors of a three-dimensional mathematically described model of the environment (Figure 15.4). The approach uses geometric descriptors such as points, lines, meshes and polygons (Foley *et al.*, 1995). The descriptors are assigned attributes such as thickness and colour. Synthetic model rendering provides a great deal of flexibility. One can move the viewpoint or alter the model in an analysis and recompute the image with a minimum of human crafting. However, one must remember that when used in practice, its lack of photographic realism can leave a lot to the imagination of the person consuming the images. Users need to be informed about how to interpret the images if the technique is used as part of a systematic working method.

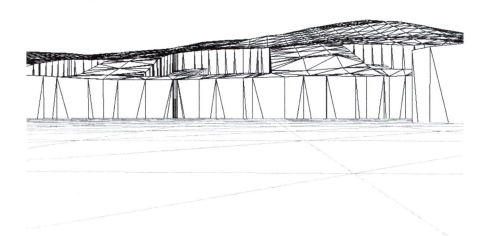

Fig. 15.4. Example of a CAD wire-frame model representing strip-harvesting at Gavin Lake in BC. Credit: John Lewis and Stephen Sheppard, Collaborative for Advanced Landscape Planning, UBC, Vancouver.

The first wireframe perspective applications of this type to VRM problems would abstract one or two key geometric parameters of the visual experience. For instance, in the 1970s the US Forest Service developed wireframe line drawing systems to plot the perspective patterning of clear cutting stands of trees on Forest Service lands (Nickerson, 1980). At CLR, we began research on this type of visualization in the early 1980s using the first generation of solid shaded computer graphic techniques developed by computer science laboratories such as the Dynamic Graphics Project at the University of Toronto. Figures 15.5 and 15.6 illustrate CLR's application of these rendering techniques on VRM issues such as transmission line visual impact and urban design view analysis.

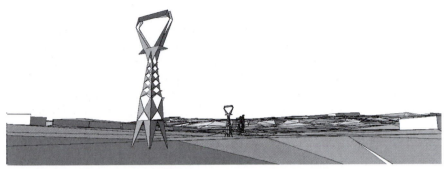

Fig. 15.5. Images from 1985 using solid shaded synthetic model imaging. The images were used in a visual impact assessment hearing on the Niagara Escarpment, Canada to illustrate the geometric parameters of electric transmission tower visibility visa via concepts explained in the accompanying written impact report. The images were used as a systematic means of explaining concepts such as "skylining the horizon" and number of towers visible. Credit: CLR, U of T, Niagara Escarpment Commission.

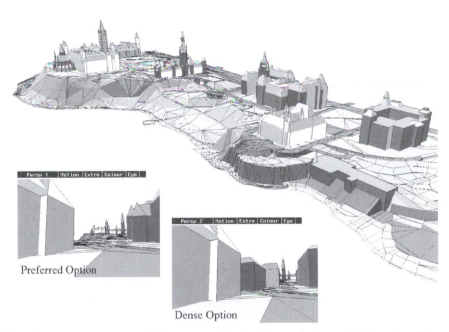

Fig. 15.6. Images from 1986 using solid shaded synthetic model imaging. The images were used to compare the visual images of various options for expanding office space in the Parliamentary Precinct Area of the Canadian Capital. Credits: CLR, U of T, duToit Allsopp Hillier, National Capital Commission.

Fig. 15.7. Illustration of temporal relationships of bat movement and built and vegetative linkages to habitat. Credits: ORL Institute, ETHZ - synthesized using CLR's PolyTRIM software.

Fig. 15.8. Illustration of overall air pollution in the Rhine Valley in the Grisons, Switzerland. Credits: ORL Institute, ETHZ, CLR, U of T - synthesized using CLR's PolyTRIM software (see Plate 13).

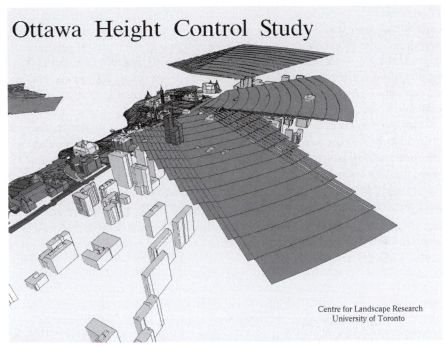

Centre for Landscape Research
University of Toronto

Fig. 15.9. Illustration of legal zoning description in Ottawa compared to actual development approvals. Credits: CLR, U of T, National Capital Commission, City of Ottawa - synthesized using CLR's PolyTRIM software.

This representation technique is extremely well suited when one wants to draw attention in a visualization to a subset of spatial parameters in the scene (as depicted in Figures 15.5 and 15.6). The approach is equally effective at systematically showing non-visual parameters of an environment in a spatial juxtaposition with visual issues. For example, zoning, and ecological factors such as habitat type, and temporal occupation of space by movements of species or plumes of pollution can be represented (Figures 15.7, 15.8, 15.9) (Danahy and Hoinkes, 1995; Hehl-Lange, 1998)

In the early 1980s, while people in VRM practice were using wireframe visualization methods, research on the production of credible fully synthetic scenes of landscape began with simulations by computer scientists and movie production companies. Images produced by Fournier and Reeves (1986) in research laboratories are typical of the possibilities demonstrated during that period. The techniques first propagated to create simulated locations for animated movie production. Until quite recently, the software, equipment and time demands of these simulation techniques were far too high to be used as a matter of course in environmental decision-making.

In theory, it is possible to imagine a visualization system that could systematically create a digital image synthesized from a simulation of visible radiation and the environment. This approach would simulate light arriving at the CCD of a digital camera that had interacted with all of the elements in a real landscape. In practice, its an uncomputable problem at the scale of a landscape and will remain so into the foreseeable future. In part, this is due to the infinite detail and complexity of landscape elements such as leaves and twigs that are not computable in any practical way by the computers available to environmental planning.

However, in the meantime, applied visualization will focus on the production of credible intermediate approximations of landscape elements. It is important for consumers of these abstractions to remember this constraint when simulations are called photorealistic. Professionals need to understand the nature of the algorithms used in computer graphics in order to understand what imagery produced with these methods does and does not portray about the landscape. Also, it is helpful to distinguish between that which is explicit, that which is approximate and that which is imaginative or conceptual.

Computation can be effectively employed in synthesized visualizations as long as the abstracted geometric descriptors of landscape elements credibly convey the information needed to assist people with a particular visual decision. A reason to focus on this type of visualization instead of image processing is that synthesized visualization is scalable with advances in computer speed. A steady increase can be witnessed in both the use and sophistication of this type of computer graphics as the speed of personal computers has increased. Therefore, it is reasonable to expect this approach to increasingly reach into everyday planning situations.

This theory was shown to be valid in examples of experimental visualization work done in the 1980's. CLR's work is an example of the application of synthetic geometric representation in real world decision-making. The results were impressive and had a significant impact on decision-making. However, the time it took to respond to questions and ideas that resulted from each iteration of image making was much too long, expensive and removed from the participants. The implication of this is that the tools could not be used on a widespread basis in day-to-day planning (Danahy and Hoinkes, 1995). Application of the techniques was limited to problems in significant landscapes for planning commissions that had a specific mandate to protect visual resources. The CLR's projects were done for cities such as Toronto and Ottawa and for agencies such as Canada's National Capital Commission, the Niagara Escarpment Commission and Ontario Hydro in the Province of Ontario.

The conclusion that emerged from these early experiments was that research is needed to examine methods that could reduce the time it took to address the iterative demands of the decision-making process. The assumption behind the effort was that professionals needed to be able to begin a basic visualization exercise within hours of receiving instructions. In addition, once the visualization was in place, people needed to be able to change viewing and model parameters as part of the dialogue in a decision-making meeting. Our overall conclusion about visualization was that if these requirements could not be met, then the techniques would continue to be reserved for special cases.

4   Convergence of Technologies

The most interesting breakthrough for landscape visualization came when image processing techniques were merged with geometric modelling. The advent of photo texturing on polygonal surface models made it possible to combine the best features of each technique. Texture mapping techniques added a new dimension to visualization that greatly enhanced the recognition of elements in a synthesized scene.

Texture mapping permits one to attach a digital image to a geometric surface. The result is an apparently photorealistic depiction of a scene with a minimum computational load. The use of sampled images of elements such as leaves and ground surfaces can efficiently convey sufficient detail in images to eliminate the potential for misinterpretation in a visualization. One can use an image on a single polygon to convey information that can take thousands of polygons to compute and represent.

Remote sensing and orthographic aerial photographs textured over terrain models made it possible to place detailed site models in a larger landscape context. The technique is computationally efficient but requires large amounts of memory to hold the bit mapped images. The widespread application of the technique did not move into practice until the cost of ram dropped. Now that 128 megabytes of memory is common, it is now very practical to exploit this potential (see Figure 15.7).

The combination of these two techniques opened the door to fully computed automation of the rendering function. This made the technique scalable and it became more viable to use in practice as technology progressed.

5   The Real-Time Visualization Breakthrough

In the late 1980s, the invention of affordable specialized perspective graphics accelerators on UNIX workstations such as those made by Silicon Graphics produced a radical change. It became possible to draw synthetic images in a fraction of a second rather than in minutes or hours as the first application experiments required. The increase in speed made it possible to research the application of these rendering techniques in a real-time interactive visualization context. The drawing speed made it possible to bring synthetic visualization into decision-making exercises and consulting. When this type of real-time computer graphics hardware became available, CLR refocused its software development in this direction to address the question of how to produce images in a more timely fashion for decision-making and dialogue.

The first generation of real-time systems were limited to wireframe line drawing images. Real-time graphics became much more useful in the late 1980s when light source shaded, solid models could be displayed. Ultimately, an added capability in the graphics hardware technology of the 1990s made it possible to

Fig. 15.10. Real-time test image of a two dimensional coverage from ArcInfo draped over a terrain model with texture mapped tree populating the coverage (1991). Credits: CLR, U of T, ESRI - synthesized using CLR's PolyTRIM software from data derived from ArcInfo.

place photo textures on geometric models with no penalty in redraw time. This meant that credible real-time interactive synthetic models could be created. The first Silicon Graphics workstations that supported this capability allowed one to place up to 512 x 512 pixels of image information into a model with no impact on the redraw time of the system (see Figure 15.10). This made it possible to show an explicit photographic representation of trees on digital terrain models. A few years later it became possible to photo texture several megabytes of images on a model with no decrease in redraw speed. This opened the door to the testing of these techniques in practice.

6   Remaining Thresholds and Barriers

These approaches were limited to smaller scale site models of one or two square kilometres due to the computational limitations of the equipment. The full viewshed context was difficult to include in these visualizations. One was usually forced

Fig. 15.11a.   Synthetic computed simulation of proposed strip-harvesting at Gavin Lake in British Columbia using World Construction Set. Credit: John Lewis and Stephen Sheppard, Collaborative for Advanced Landscape Planning, UBC, Vancouver (see Plate 14 and 15).

Fig. 15.11b. Image derived from the automatic assembly of GIS, Remote Sensing, CAD and database information: a planning workshop for the Tennesse Valley Authority by Steinitz, Ervin and Hoinkes. Credit: Harvard GSD, CLR, U of T - synthesized using CLR's PolyTRIM software (1994).

to severely abstracting background terrain and draped image data. Vegetation was severely limited and could seldom explicitily depict a representation of every tree on a hillside. This limitation is being solved as computers become ever more powerful each year. Figure 15.1 illustrates a scene of Emerald Bay on Lake Tahoe. The image illustrates what curatorial historians considered just enough elements of terrain, rock outcropping, architecture and vegetation cover to be considered usable to convey the site planning intent of Frank Lloyd Wright's design (DeLong, 1996).

The power of personal computers and sophisticated software now costs one or two orders of magnitude less than the simulation systems of a decade ago. Single frame visualization of landscape with background vegetation models is practical to consider in situations where a good digital database is available. On many levels, the images one can make with software that costs less than one thousand dollars provide more realistic looking foreground detail and provide more credible representations of distant background landscape than any previously available visualization software (see Figure 15.11a).

In addition to the improvements in computer graphics discussed in this chapter, data creation for visualization systems has been a time consuming barrier to the application of visualization in practice. Computational automation has been used to eliminate this constraint in applied research applications (see Figure 15.12) (Hoinkes and Lange, 1995). The latest commercial software systems incorporate these concepts and make it more practical to systematically access base data

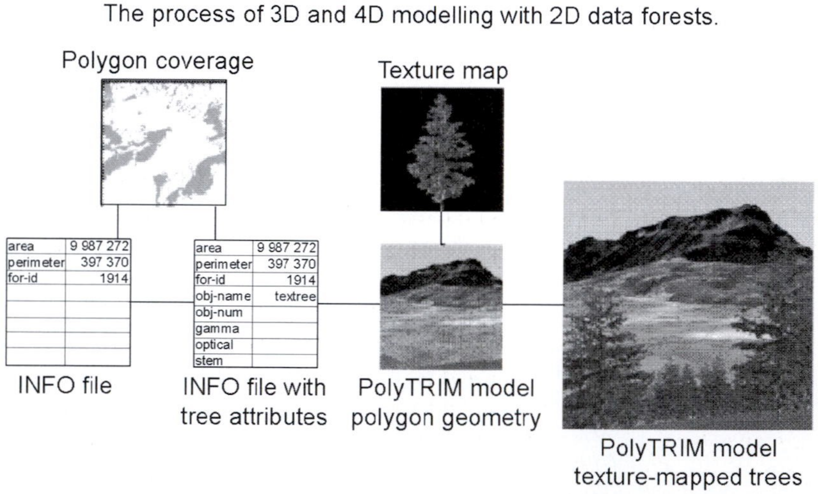

Fig. 15.12. Image derived from the automatic assembly of GIS, Remote Sensing, CAD and database information developed by R. Hoinkes and E. Lange. Credit: ORL Institute, ETHZ, CLR, U of T - synthesized using CLR's PolyTRIM software (1994).

prepared in a GIS and database systems used for day-to-day planning. Creation of both the visualization imagery and the database it needs can benefit from automation techniques.

Furthermore, increasing functionality for the translation between computer graphic data structures and the data structures of other software systems makes it more practical to link visualizations to simulation models, such as ecological or tree growth models. The ability to systematically produce credible visualizations of predicted or alternative futures derived from simulations of natural and human processes will be a profound amplification of decision-making and landscape management practice.

As more visualizations are systematically and automatically derived from online data bases of a site, a new and equally valuable role for visualization is likely to emerge. Visualizations will be used as a visual-spatial search interface to online information and knowledge about the site and the underlying knowledge used to create an ecological simulation. It is technologically possible to query the pixels of an image depicting a tree or forest block in a photo textured visualization to receive information and explanations about matters such as the ecological status of the element. When this functionality is available, visualization should occupy a more ubiquitous role in all aspects and stages of environmental decision-making. Visualization will not simply be reserved for simulations of virtual visual sensory experience. Increasingly, we expect visualization will be used as a primary interface to spatially referenced online information (see Figure 15.13).

7 Visualization Data Abstraction Considerations

The first part of this chapter positioned the current use of visualization techniques in the context of the evolution of the technology. It made the point that the state-of-the-art in visualizing landscapes has only just crossed the combined thresholds of representative credibility, cost and timeliness in decision-making dialogue. However, one cannot take the computational complexity of visualizing landscape for granted. As such it is still necessary to strike a balance between computational demand and the realism of images.

In this section, a matrix of abstraction considerations and heuristics used at the Centre for Landscape Research to guide the production of interactive visualization projects is presented. It identifies and relates a series of parameters that influence the effectiveness and cost of interactive real-time visualizations. The Centre for Landscape Research has sought to consult with cities and public agencies that share an interest in applying digital visualization technology to solve problems conventional professional practice and commercial software fails to address. Our focus is on the demands of fostering visual-spatial dialogue in the design and decision-making process. As such, CLR's emphasis has been on exploring the possibilities of high-speed interactive graphics for visualizations of landscape. This approach required that we carefully tune representations and models to be as computationally simple as possible yet professionally credible.

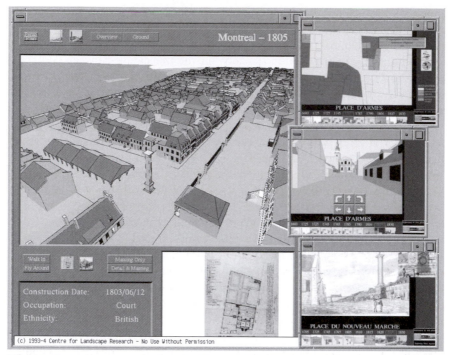

Fig. 15.13. An example of a visualization model being used as the primary interface to querying of an online database. Image illustrates a real-time interactive timeline model of old Montreal. The visualization is built in real-time as one move the timeline slider. Inversely, one can select elements in the image and receive information from the database. Joint research project between CLR and CCA. Credits: CLR, U of T, Canadian Centre for Architecture. Synthesized using CLR's PolyTRIM software.

CLR has incrementally developed a research testbed software system designed to take advantage of specialized computer graphics hardware. Research has focused on ways to use the fastest rendering tools. Techniques that diminish the speed of drawing images without returning a significant transformation in visual information relevant to a decision-making process have been avoided. The system created by CLR is called PolyTRIM. It is a toolkit system developed on Silicon Graphics computers. It combines data types and computational tools found in a range of digital design and planning media (virtual reality, GIS, Computer Aided Design (CAD), animation, sound, hyper-links, collaborative-work via Internet, image processing, text, digital library tools, database, interactive exhibitions, customizable interfaces, user history tracking, scripting language and 'C' library support) (Danahy and Hoinkes, 1995). The idea behind the testbed is to operate as a meta visualization tool that works on top of conventional technologies such as data base, computer animation, imaging, rendering, design, and geographic information systems. The images used to illustrate this chapter were created with PolyTRIM.

The computational resources available for any given visualization exercise are always finite. The time available to complete a visual resource management exercise is finite. These two factors should remain in the forefront of one's consciousness throughout the visualization process. There is never enough computer power and there is never enough time to satiate peoples' curiosity during an active and vital decision-making process. Compute power and time govern the choices one must make in selecting the appropriate balance between image and geometric detail. No matter how much memory and speed our computers attain, synthetic virtual reality models will only be able to approximate the detail and intricacy of a real landscape. Therefore, one must recognize this parameter of the problem at the outset of any visualization exercise. Additionally, one must judge whether the degree of abstraction necessitated by a given visualization technique is acceptable.

Another issue worth examination in designing a visualization exercise is the specification of the duration of the feedback loop. This refers to the amount of time that can elapse between when the need for a particular view or change to the virtual description of the synthetic environment is identified and the time that the visualization is available for the decision-maker to evaluate. This variable will influence the choice of hardware and software. For example, if requests to change the vantage point in a simulation can wait until the next meeting, then one can make use of a typical office or laptop computer. On the other hand, if people's time is scarce and it is logistically difficult to reassemble the group one will need to consider adding a specialized graphics card with a large texture memory. Also, the database will need to be structured to ensure that the computation can all be done in the random access memory of the computer without resorting to swapping sections of the data onto the hard drive during the computation. If this happens the process becomes exceedingly slow and will not be practical to use during a meeting.

The most important factor governing the selection of a visualization technique is the identification of the specific models of knowledge one is attempting to communicate or replicate in a simulation. In the case of forest landscapes, ecological and aesthetic models need to be linked to the structure of the visualization (see Chapters 3 and 11 for examples). For instance, some models of aesthetics such as symbolic aesthetics may not be well served by the exclusive use of photorealistic visualization techniques since a particular concept of beauty may require the association of non-visual knowledge with the representation of a scene (Lange, 1999).

Most computer graphics systems focus on the problems and interface vocabulary of computer graphics algorithms and data structures. The basic spatial vocabularies of these systems centre on terms such as polygons or quadratic surfaces. An application such as environmental planning requires a vocabulary used by the thought processes of the end users. This means terminology needs to be problem based. The computational operators supported by a visualization programme should in some way amplify the task in a structure that is part of the discipline using the system. The computer graphics terminology of most visualization software can act as a literacy and thought barrier.

In the case of landscape representation, CLR's clients have consistently led us to use a vocabulary with the following three types as a minimum lexicon of elements in base models. The three most common sub types needed in a base model are terrain, vegetation, and built form. Water has served as a fourth element in some settings such as lakes or oceans. We find that water is seldom manipulated as part of the decision-making issue except when a dam is being tested. The most fundamental element is terrain since all vegetation and built form are situated relative to the terrain in reality. Even water is contained by an underlying terrain. The clients we have served all needed the images and databases to convey this morphological reality if the visualizations were to be credible and provide information about spatial relationships that were too complicated to portray with manual representation methods (see Figure 15.1 and Plate 9).

Determining the level of detail is a necessary consideration in using the computational budget efficiently. Viewpoint is one of the most important determinants in selecting the appropriate level of detail needed for display by a computer graphics visualization system. One needs to predict approximately how much visually recognizable detail will reach an observer's eye. If the viewpoint is a long distance away from a tree it is very unlikely that a person could discern much more than a typological understanding. For example, one may see the object as a pine tree but not be able to discern which species of pine. If the viewpoint is far enough away, it may appear more like a symbol of a tree with it only registering or stimulating a very small number of light receptors (cones) in the retina of a hypothetical viewer. One does not need to use a complex, explicit representation (in either geometric or image detail) to produce a useful and credible image as landscape elements get farther away from the vantage point. Conversely, if one is within 30 metres of a tree, one is going to begin to need both three-dimensional geometric spatial information and quite specific image detail for elements such as trees. Also, if the trees have a distinctive visual character from one another, then one will need to use exact texture maps of the real trees (explicit detail). If the shape of each tree is unique, then the geometric description of each tree may also need to be captured before the final visualization will be seen as credible.

The simplest approach taken to this issue by visualization systems is to store multiple levels of detail in memory and use distance from the viewpoint algorithms to send the optimum balance of data to the rendering pipeline. The convention of foreground, middle ground and background levels of detail is often used. This strategy allows one to balance the computational load on the system based on viewpoint position in the model.

We have found it extremely useful to structure the geometric models on at least three basic levels of detail (symbolic, typological and explicit). Images used for texture mapping are equally sensitive to issues of abstraction and resolution. If used in an unstructured way, texture mapping can become as time consuming

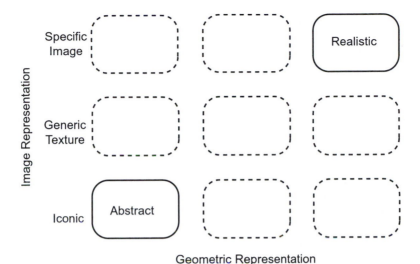

Fig. 15.14. A matrix of representation possibilities: Illustrates the types of combinations of levels of representation detail surfaces can have using typical real-time computer visualization systems.

and computationally difficult to use efficiently as geometric representation. The same principles of abstraction and level of detail can be applied to this technique. We categorize textures in terms of icons, generic images, and specific images of particular (real) things in a site.

The matrix shown in Figure 15.14 is presented as a guide for people to "map" and consider the implications of alternative combinations of abstraction. The key to making projects work lies in the development of an abstraction strategy that properly represents the primary elements of the landscape being studied. This applies equally to making images that are to be judged as credible representation of what the human visual system would sense and to making images that present a visual emphasis to communicate an abstraction or interpretation. The approach described above forms a three by three matrix of representation possibilities.

8   Conclusion

Users of visualization systems today have an extremely rich range of representation options at their disposal. The apparent photo-realism of static synthesized views and frame-by-frame animation for visualization of large-scale forestry practices has crossed new thresholds of usefulness. During the last part of the 1990s, commercial software such as World Construction Set sufficiently adapted its algorithm and data types to real-world representation needs to make it possible to synthesize images of large scale forested landscapes in a systematic and parametric way. If these large scale, high detail rendering capabilities can be paired with highly interactive software visualization techniques that take advantage of accelerated computer graphics hardware and link other forms of site information to visualizations of both types then we should enhance our capacity to systematically associate abstract knowledge representations with visually explicit representations. The software prototypes developed in the 1990s at research labs such as the Centre for Geographic Information Systems and Modelling at University of Melbourne, the Imaging Systems Laboratory at the University of Illinois and the CLR at the University of Toronto.

The projects undertaken at CLR have shown to us that the most interesting applications of visualization cannot be readily exploited in everyday practice without carefully structured and abstracted data. Further, the costs of assembling and crafting data through interactive modelling are too high for most day-to-day planning. Continued attention to automating the creation of geometric base data will solve much of this problem. The gathering of vertical surface photo textures of specific trees and buildings remains the most significant barrier in the problem. Basic research in photogrammetric methods to automatically extract buildings from aerial images is making steady progress (Danahy, 1999). In general, more work is needed on the automation extraction of textures and geometry from both aerial and terrestrial sensed images.

Our experience with clients has been that as they become more involved in the process they begin to treat the elements of a visualization less as a finished picture or information commodity. They begin to develop a literacy with the vocabulary of the model and the computational operators if they can explore the model for themselves. Users of a robust system can enter into a creative process of mistake making. Ideas and hunches can be explored in seconds. As familiarity builds and when the imagery directly reflects the vocabulary of the decision process, we find people begin to use abstract visualization integrally with conventional photorealistic visualization. The medium makes it very simple to compare representations and maintain a systematic spatial reference. The range of visualization options from abstract to explicit get used as a language to question, debate and create possible futures.

The process will become a matrix of representation possibilities with one systematized medium instead of the dichotomy that has existed between abstract and explicit imagery.

References

Danahy, J.W. (1987) Sophisticated image rendering in design. In: *The Association for Computing Machinery Proceedings of SIG CHI+GI '87*, Toronto.

Danahy, J.W. (1999) Visualization data needs in urban environmental planning and design. In: W. Fritsch/Spiller (ed) *Photogrammetric Week 99* Stuttgart.

Danahy, J.W. and R. Hoinkes (1995) *PolyTRIM*: Collaborative setting for environmental design. In: M. Tam and R. Teh (eds) *CAAD Futures '95,*

The Global Design Studio. <http://www.clr.utoronto.ca/PAPERS/CAAD95/caadf.jd8.html>.

DeLong, D.G. (1996) *Frank Lloyd Wright: Designs for an American Landscape, 1922-1932.* Thames and Hudson, London.

Foley, J., A. van Dam, S. Feiner, and J. Hughes (1995) *Computer Graphics: Principles and Practice,* Second Edition. Addison-Wesley, Reading.

Fournier, A., and W. Reeves, W. (1986) A simple model of ocean waves. In: *Proceedings of the 13th annual ACM Conference on Computer Graphics,* pp. 75-78. ACM, Dallas.

Gruen, A., E.P. Baltsavias and O. Henricsson (eds) (1997) *Automatic Extraction of Man-Made Objects from Aerial and Space Images (II).* Birkhauser Verlag, Basel.

Hehl-Lange, S. (1998) Funktionen und Wirkungen von Lebensraumtypen und deren Bedeutung für die Ökologischc Planung - GIS-gestützte Analyse und Visualisierungeines potentiellen Fledermaus-Jagdhabitats. *Natur und Landschaft,* 73, 256-260. <http://www.orl.arch.ethz.ch/~Lange/goldau/goldau.html>.

Hoinkes, R. and E. Lange (1995) 3D for free - Toolkit expands visual dimensions in GIS. *GIS World,* July.

Lange, E. (1999) Realität und computergestützte visuelle Simulation. Eine empirische Untersuchung über den Realitätsgrad virtueller Landschaften am Beispiel des Talraums Brunnen / Schwyz. ORL-Berichte Nr.106, VDF, Zürich, 176 S. <http://www.orl.arch.ethz.ch/~Lange/schwyz/schwyz.html> and <http://www.orl.arch.ethz.ch/~Lange/cottbus/cottbus.html>.

Nickerson, D.B. (1980) *Perspective Plot: An Interactive Analytical Technique for the Visual Modelling of Land Management Activities.* In-service Report. USDA Forest Service, Portland, OR.

Orland, B. (1993) Synthetic landscapes: A review of video-imaging applications in environmental perception research, planning, and design. In: D. Stokols and R. Marans (eds) *Environmental Simulation: Research and Policy Issues,* pp. 213-252. Plenum Press, New York.

Sheppard, S.R.J. (1989) *Visual Simulation - A User's Guide for Architects, Engineers, and Planners.* Van Nostrand Reinhold, New York.

Sheppard, S.R.J. (2000) Visualization as a decision-support tool for managing forest ecosystems. *COMPAG 27* (1-3).

Chapter sixteen:

Predicting Preferences for Scenic Landscapes using Computer Simulations

JoAnna Ruth Wherrett

Land Use Science Group, Macaulay Land Use Research Institute, Craigiebuckler Aberdeen

1   Introduction

One approach to the problem of evaluating scenic resources is to produce a mathematical model which can predict the general public's aesthetic preferences for a number of landscapes, by assessing the components within the landscapes. Such models can be used to predict relative changes in landscape preference, as a landscape changes or under different management scenarios. The methodology may also be extended to explore other perceptions of the public, such as sustainability.

A visualization of the landscape is required to create predictive preference models. These landscape visualizations have traditionally been slides, photographs or scanned photographs seen on a monitor. However, if computer simulations of landscapes were able to produce similar aesthetic responses to photographs of landscapes, it would be possible to use such simulations to create preference models. This would also have the advantage that the landscapes rendered would be geographically referenced, and that any landscape component data required may be obtained from data held within a geographical information system (GIS). By linking such models with a GIS, interactions between indicators of sustainability and aesthetic perception may be explored.

This paper will examine two questions: firstly, do photographs and computer simulations of the same landscape produce similar aesthetic quality judgements from the general public; secondly, do landscape preference models, created using photographic images, also work with the landscape component data from computer simulations. To answer these two questions, an experiment has been undertaken into scenic perceptions of a variety of landscape simulations types. Before detailing the experiment, the landscape preference models which will be used to compare predicted results to the preferences of the general public are described, with both data collection and analysis detailed.

2 Psychophysical Landscape Preference Models

Psychophysics is the study of measurement theory and procedure which attempts to relate environmental stimuli to human sensations, perceptions and judgements (Hull *et al.*, 1984).

Psychophysical preference modelling is one of the quantitative holistic techniques of landscape evaluation which mix subjective and objective methods. The relationships of interest in such modelling are those between the physical features of the environment, such as topography, vegetation and water, and the psychological responses, which are often judgements of preference, aesthetic value or scenic beauty. To develop such models, a range of different landscapes must be assessed and their physical characteristics measured (see, for example, Shafer *et al.*, 1969; Bishop and Hulse, 1994).

Of all the methods of landscape evaluation, the psychophysical models have been subjected to the most rigorous and extensive testing. The models have been shown to be very sensitive to subtle variations in landscape and have proven very robust to changes in landscapes and in observers (Daniel and Vining, 1983). However, there are several drawbacks to the use of psychophysical models: they require the full range of possible scenes to be selected to represent all of the physical characteristics to be used as predictors of scenic beauty (Hull and Revell, 1989) and they can be expensive and time consuming to develop (Daniel and Vining, 1983).

The following section will detail the collection of the psychological and the physical data used to create the preference models for this study. These models are then used later in the research to predict preference scores given to a variety of landscape simulations.

2.1 Psychological Data - The Landscape Preference Questionnaire

The method adopted here for landscape preference data gathering was a computer-based approach using the graphical user interface of the Internet, *i.e.* the World Wide Web (WWW). This method of conducting a preference questionnaire is a novel approach to the problem of collecting a sufficiently large sample of responses to photographic stimuli (landscape images) for the creation of a predictive preference model. Responses for both the surveys described were gained by e-mailing a variety of address lists and mailbase groups, responses also came from a variety of people who found the WWW site through search engines. It is acknowledged that using such a sample group may produce a bias in the results. Research into this question conducted by Wherrett (1998) showed that any such bias was minimal for the data used in the research reported here.

The landscape preference questionnaire ran on WWW browsers, using hypertext markup language (HTML) documents and PERL (Practical Export and Reporting Language) scripts. In general terms, the respondents read an introductory page, were then given instructions and asked to proceed to the questionnaire appropriate for their computer screen size. They were then asked to practise scoring using a single landscape image. At this point the respondent was assigned a random number corresponding to a particular set of photographs. They were then given ten photographs to look at before progressing to a scoring page containing the same ten photographs. A second set of ten photographs was then browsed and scored, and finally some socio-demographic information was requested. Respondents were asked to score each photograph between 1 (low scenic preference) and 7 (high scenic preference).

Ninety photographs were used in the survey, representing a wide range of landscape types in Scotland, with a wide variety of topographies and land covers, from different geographical areas. These were then randomly placed into 9 groups of 20, each image being seen in two different groups. Respondents were the assigned one of the 9 groups at random. A total of 180 responses were gathered over a four month period in 1997. These scores were then normalized in two ways: firstly per questionnaire, and secondly per respondent. The equations used are detailed in Wherrett (2000).

2.2 Physical Data - Landscape Component Variables

Landform Variables

The work of Brush (1981) suggests the use of landform as a preference predictor. Landscapes were divided up into the five types of landform: flat (LF/PLF); low hill (LL/PLL); steep hill (LS/PLS); mountain (LM/PLM); and obscuring vegetation (OV/POV). Definitions of the landform types can be found in Wherrett (2000). Sky (SKY/PSKY) was also used as a landscape variable, although it relates more to photograph compostion than landscape preference (see Hammitt *et al.*, 1984). Water (WATER/PWATER) was used as a landscape variable and was also divided into three types: still, moving and sea water (e.g. AWSTILL, PWSEA). The area and perimeter of each type was calculated by digitizing the photographs in ERDAS Imagine into polygons and using the software to measure the size of each polygon.

Colour Variables

Colour variables are objective characteristics of the images and are measured using algorithms available within ERDAS Imagine. Within an IMG raster file each pixel is assigned a level for red, green and blue bands between 0 and 255. The mean, standard deviation and variance for each colour band were calculated giving six variables for each photograph. REDVAR therefore is the value of the red band variance.

Cognitive Proxy Variables: Bytes per Pixel

The link between complexity and preference was first suggested in research by Berlyne (1963) and Wohlwill (1968). The measurement of the complexity variable was suggested in the works of Bishop and Leahy (1989) and Orland *et al.* (1995). Their studies used a variety of methods to calculate complexity of the image, one of which was to examine the compressibility of the image. GIF format uses lossless compression, reducing file size by recording the length of a horizontal line of the same colour instead of recording the colour of each pixel in that line (Bishop, K., 1997). One definition of complexity that could be used is proximity of colour changes within an image. Therefore, the file size of the GIF image represents the level of compression available for each image thus corresponding to the level of complexity of the image. To identify the number of bytes per pixel in the original GIF image, the size in pixels was recorded and the size in bytes was divided by this figure. The notation for this term is BPP.

Mathematical Transformations

Mathematical transformations were made of each variable. The inverse, natural logarithm, ratio of area to perimeter and square of each variable were calculated. These transformations were chosen as they are the most likely relationships to be appropriate measures of natural phenomena.

In the cases of the inverse, natural logarithm and ratio divisor, the value of 1 was added to each variable value to prevent errors occurring when the variable value was equal to zero. The exceptions to this were the area and perimeter of sky and bytes per pixel, none of which may ever equal zero. In several cases the ratio of two different variables was calculated (*e.g.* R_PSKYPWATER).

It should be noted that variables including measurements of sky are useful only in predicting preferences for images of landscapes and have no use in a managerial sense.

2.3 Landscape Preference Models

A number of psychophysical landscape preference models were created using multiple linear regression with forward variable input on the preference data and landscape component analysis data; further details of the methodology used can be seen in Wherrett (2000). Three of these models (Tables 16.1, 16.2 and 16.3) were used in this study. Models A1 and A2 are broadly similar, sharing three variables. Model F is slightly more complex, but uses no colour or cognitive variables and can therefore be used without the need for the images to be digitally processed.

If these models are examined closely, it can be seen that several variables occur within all three models. The inverse of the perimeter of mountain landform (INV_PLM) term and the inverse of water/ratio of the perimeters of sky to water (INV_WATER/R_PSKYPWATER) terms are common to all three models,

Table 16.1. Model A1: Regression output.

Model A1	R^2 = 0.658 (adjusted = 0.638)		Normalization: per questionnaire		
Variables	Unstandardized Coefficients		Standardized Coefficients		
	B	Standard Error	Beta	t	Significance
(constant)	0.322	0.100		3.236	0.002
INV_WATER	-0.324	0.045	-0.508	-7.244	0.000
INV_BPP	-123.511	25.135	-0.352	-4.914	0.000
INV_PLM	-0.247	0.040	-0.411	-6.247	0.000
REDVAR	4.249 E-5	0.000	0.241	3.456	0.001
LN_LS	2.120 E-2	0.010	0.151	2.220	0.029

Table 16.2. Model A2: Regression output.

Model A2	R^2 = 0.642 (adjusted = 0.625)		Normalization: per questionnaire		
Variables	Unstandardized Coefficients		Standardized Coefficients		
	B	Standard Error	Beta	t	Significance
(constant)	0.406	0.090		4.486	0.000
INV_BPP	-124.565	25.436	-0.355	-4.898	0.000
INV_PLM	-0.252	0.040	-0.420	-6.359	0.000
R_PSKYPWATER	-2.82 E-3	0.000	-0.479	-6.889	0.000
REDVAR	3.923 E-5	0.000	0.223	3.196	0.002

reflecting the importance of water and mountainous landform to landscape aesthetics. In both cases the form of the variable (inverse) implies that it is the presence or absence of these terms which is most important rather than their absolute value.

The negative contribution of the complexity term (INV_BPP) appears in Models A1 and A2, implying that complexity is a positive predictor of landscape preference, with more simple, less diverse landscapes likely to have lower predicted scores than more varied landscapes. The red band variance term (REDVAR) shows a positive contribution to landscape preference. Larger variance in the levels of red within a landscape may signify more variety or more difference in tone within the photograph.

The third model has a far larger range of variables than Models A1 and A2. When some of the terms are examined together, some interesting effects appear. The three types of water do not have equal effects on predicted scenic preference; still water appears to be more aesthetically pleasing than moving water or sea water. The lower landform variables, flat land (LF) and low hill (LL) both have negative contributions to scenic preference, however, if both are present the magnitude of the term is smaller than if only one landform is present. Finally, if the sky terms are examined, it can be seen that the positive contribution of sky peaks at around one quarter of the photograph area.

Table 16.3. Model F: regression output.

Model F	$R^2 = 0.724$ (adjusted $= 0.681$)		Normalization: per respondent		
Variables	Unstandardized Coefficients		Standardized Coefficients		
	B	Standard Error	Beta	t	Significance
(constant)	-2.300	0.869		-2.648	0.010
INV_WATER	-0.491	0.094	-0.420	-5.227	0.000
INV_PLM	-0.374	0.083	-0.341	-4.514	0.000
PWSTILL	3.642 E-3	0.001	0.332	3.981	0.000
SKY	-2.52 E-3	0.001	-0.692	-3.544	0.001
LN_SKY	0.620	0.186	0.631	3.330	0.001
R_WSEA	0.310	0.088	0.422	3.504	0.001
INV_PLL	-0.266	0.091	-0.239	-2.906	0.005
OBSVEG	7.156 E-4	0.000	0.249	2.611	0.011
FL^2	-1.23 E-6	0.000	-0.196	-2.373	0.020
INV_LF	-0.193	0.085	-0.183	-2.267	0.026
(LS+LM)	5.953 E-4	0.000	0.186	2.147	0.035
$PWSEA^2$	-3.45 E-5	0.000	-0.246	-1.993	0.050

3 Experimental Study into Aesthetic Preference of Landscape Visualizations

Many simulations of natural landscapes are used in the planning process, often to show the general public what a proposed development or change may look like. While many of the images are geometrically accurate, it is not known if they are valid to be used for landscape preference studies.

It is important to understand what the criteria for realistic simulations (*i.e.* those which produce the same perceptual response as real landscapes) of landscapes are; without this knowledge all visualization methodologies would need to be validated before they are applied to landscape proposals. This part of the study looks at the latter point, and examines the photo-realism and perceptual realism of three different methods for landscape visualization. It is accepted by many authors (Dunn, 1976; Shuttleworth, 1980; Stamps, 1990) that photographic images of landscapes are perceived in essentially the same manner as the actual landscape. The following survey examines the hypothesis that people will rank a set of simulated images in the same order as a set of photographic images.

3.1 Methodology

A site was chosen for a comparison between landscape simulations and actual landscape photographs around Braemar in north-east Scotland; the site was chosen for its accessibility and the availability of large and detailed datasets at the Macaulay Land Use Research Institute in Aberdeen. A digital elevation model

(DEM) at 5 metre resolution exists for an area of approximately 20 square kilometres to the west of Braemar and colour orthophotography also exists for the entire of this area, at 2 metre resolution.

Three techniques were used to simulate the landscapes.

1. EDRAS Imagine Perspective Viewer;
2. Vistapro(tm) (Hinkley, 1997) (using aerial imagery as ground colour); and
3. Vistapro(tm) (using automated terrain coloration and manual object placement).

The sites were extensively photographed in the summer of 1998 from all positions which were viable for landscape simulation, that is to have complete horizon views. Skies were not required to be simulated due to the cloudless weather conditions on the fieldwork days. The simulated models were then manipulated to produce snapshots of identical scenes. For the ERDAS Imagine and Vistapro using aerial imagery scenes, the viewer was placed at the same grid reference and direction as the photograph was taken and the scene rendered. Due to the higher accuracy of grid references in the software compared to that on the ground, the viewer was then moved and the direction altered until the rendered scene matched the photograph. The range of visible mountain and hill horizons was used significantly in this process. For the Vistapro with terrain coloration and manual placement scenes, the same grid references were used as for the alternate Vistapro scenes. Trees and rivers were then digitized into the scene, using the photograph and the aerial imagery as a reference.

Ten of these scenes were then input into four sets of questionnaires for preference scoring, using the same Internet survey methodology as previously employed. Each questionnaire version used ten images created using the same software, which were then compared to one questionnaire version using the photographic images.

Analysis of the simulation images was undertaken in the same manner as for photographs. All colour variables were measured and the photographs and simulations were digitized using ERDAS Imagine (ERDAS, 1994).

The following example (Figure 16.1; Plate 16) shows all four methods of landscape simulation. The view is taken from the Linn of Corriemulzie towards Creag Bhalg, in the area west of Braemar, Aberdeenshire.

3.2 Results

Question One: Do photographs and computer simulations produce similar aesthetic quality judgements?

A total of 89 replies were received; their distribution between the four questionnaire versions and corresponding Intergroup Reliability is shown in Table 16.4. The reliability level was calculated by sub-dividing each group into two and

254 Wherrett

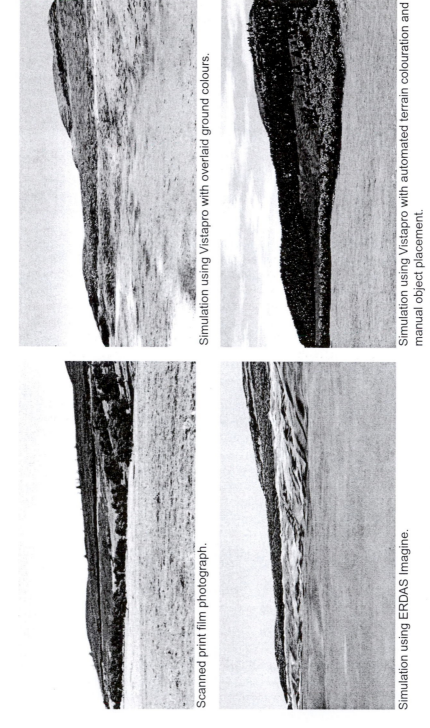

Simulation using Vistapro with overlaid ground colours.

Simulation using Vistapro with automated terrain colouration and manual object placement.

Scanned print film photograph.

Simulation using ERDAS Imagine.

Fig. 16.1. Landscape visualizations, Linn of Corriemulzie (also see Plate 16).

calculating the correlation coefficient between the averages of the two sub-groups. This process is repeated several times, with random sub-groups, to produce an overall level of reliability.

The reliability of the results for the photographs is higher than that for any of the visualizations (Table 16.4). This is not unexpected, as visualized landscapes

Table 16.4. Intergroup reliability.

Visualization Method	Intergroup Reliability	Sample Size
Photograph	0.930	23
ERDAS Imagine	0.663	20
Vistapro (aerial imagery)	0.864	21
Vistapro (manual placement)	0.731	25

may require a far greater degree of interpretation by the viewer than standard photographic images. Nevertheless, the intergroup reliabilities are still significant, and do not reflect a lack of consensus among the respondents.

The correlations between the preference scores and ranks from the questionnaire survey of the ten photographic images and the three types of visualized images are shown in Table 16.5. At best these correlations are modest, with the Vistapro (manual placement) images the only set to achieve significance at the 0.1 level. These results suggest that the images are not being scored in the same manner as the photographs; it is hypothesized that while a respondent may be able to see a photograph and 'translate' that image into a level of aesthetic quality, it is not possible for them to 'translate' a simulated image into a 'real' image in order to give the landscape an aesthetic value.

Table 16.5. Correlations between preferences for photographic and simulated images (*significant at p. ≤ 0.1 level).

	Normalized Results	
Visualization Method	Scores	Ranks
ERDAS Imagine	0.338	0.285
Vistapro (aerial imagery)	0.369	0.333
Vistapro (manual placement)	0.635*	0.588*

Question Two: Do predictive landscape preference models work with landscape component data from computer simulations?

Table 16.6 shows correlations between actual and predicted preference scores for the photographs and the Vistapro (manual placement) simulations. The predicted

Table 16.6. Correlation between actual and predicted results (**significant at p ≤ 0.05 level, *** significant at p ≤ 0.01 level).

Model	Photograph		Vistapro (manual placement)	
	Score	Rank	Score	Rank
Model A1	0.749**	0.818***	0.202	0.107
Model A2	0.768***	0.794***	0.180	0.179
Model F	0.632**	0.770***	0.223	-0.466

scores were calculated by inputing the landscape component variables from the photographs and Vistapro simulations respectively into the three psychophysical models detailed earlier.

The correlation for the photographs are uniformly high showing that the models produced reasonably reliable predictions for photographic images, while the correlations between the Vistapro images are, in the majority, very poor.

Table 16.7 details the correlations between the scores given to the photographs and the predicted scores created using the landscape component data from two of the simulation types. These results will show whether or not the simulations may be used to derive the input to landscape preference models.

Table 16.7. Correlations between predicted (Imagine and VIstapro) and actual (photographs) results (**significant at p ≤ 0.05 level, ***significant at p ≤ 0.01 level).

Model	ERDAS Imagine		Vistapro (manual placement)	
	Score	Rank	Score	Rank
Model A1	-0.913***	-1.000***	0.911***	0.929***
Model A2	-0.939***	-1.000***	0.901***	0.857**
Model F	0.220	0.079	0.755**	0.855**

The results show that the Vistapro simulations are able to predict the actual preferences of respondents for the photographs whereas the landscape variable data from the ERDAS Imagine simulations are unable to predict these preferences (no significant positive correlations).

The Vistapro (manual placement) results show significant correlations with the actual photograph preference; these compare very well with the correlations between the actual and predicted scores of the photographs. The ERDAS Imagine simulations predict preference scores which are opposite to the respondents' preference scores for the photographs - the ranks for the ERDAS simulations are a perfect negative correlation. This result, while statistically significant, is unlikely to have any meaning other than that the simulations are not accurate enough to be able to provide landscape variable input similar to that of the original photograph.

One possible reason for this is that the ERDAS simulations are missing some of the finer detail which it is possible to achieve in the Vistapro (manual placement) simulations.

4   Discussion

The empirical studies detailed in this paper attempted to answer two questions: do people perceive photographic and simulated landscapes in the same way, and can the landscape component data within a computer simulation of landscape be used to predict landscape preference.

The first part of the study showed that people do not respond to computer simulations of landscapes in the same way that they respond to photographic images. This finding is backed up by the work of Daniel and Orland (1997) who also found low consistency in ratings of scenic beauty for abstract management views. They noted that such views were "insufficient for the purpose of obtaining public judgements of perceived scenic beauty". The work of Lange (1998) has suggested that terrain models with aerial photography overlaid can produce images of very high perceived realism, when only the background is viewed. It is possible that the use of all distance zones has decreased the level of realism in the simulated images to a point where they cannot be perceived as a realistic landscape image (Lange, 1998).

At this point it is worth considering the roles of such visualizations? Do we wish to show these images to the general public in order to gain information about their preferences, or do we wish to use them in a management driven way. Perhaps the question that should be asked is: can predictive preference models, applied to visualised images of landscapes, predict preferences for the landscapes those images are simulating?

The second section of the research compared predicted aesthetic preference, using computer simulations for landscape component data input, with actual preference scores for both simulated and photographic images. Results showed that while the simulations did not predict the scores which they were given, some simulation types were able to predict the preference scores given to the photographs. This result may reflect the nature of the visualization software used; the non-dedicated landscape visualization software is unable to predict preferences for simulated images. The dedicated software (Vistapro) is able to simulate landscapes to the degree of realism necessary to be able to use predictive preference models.

If the form of the two types of simulation are examined (DEM with draped imagery and object placement techniques), it can clearly be seen that the technique using placement of objects is closer to reality in the ability to judge elevation and obscuring vegetation. However, there is also an artifact of the method of simulation in the results; the ERDAS Imagine software used the data available to create visualizations of the landscape from the same position as the photograph was taken from. The Vistapro software improved on this by allowing the landscape and objects within it to be manipulated to match the photographs as closely as possible. Nevertheless, this does not explain lack of correlation between the predicted results from the Imagine visualizations and the photographs.

5 Conclusions and Future Research

This research has only looked at three fairly simple methods of landscape simulation, and has shown that while none of these methods may be used for gathering preference data from respondents, one of the methods may be used for predicting landscape preference for current and alternative future landscapes. In order to use this information for assessing which visualization methods are most likely to produce photorealistic images, a selection of variables describing the simulation would need to be found. Such variables could then be compared to a wider variety of simulation techniques in order to assess the criteria for perceptual realism.

The results lead us to several questions about the future for simulations in landscape preference prediction. The use of simulations may lead to increased efficiency in use of predictive models, through the automation of the calculation of landscape variables and the use of process-based environmental models to derive landscape changes. However, simulations may never be able to fully convey the information that is contained within a photograph to an observer, despite increasing levels of photo-realism. For systems with some finer detail, such as SmartForest (Imaging Systems Laboratory, 1995) or MONSU (Pukkala, 1999), the simulations produced should be able to drive preference models, created from photographs, with a high degree of success. These geo-referenced forest resource models would then offer potential for testing and comparing perceived levels of aesthetic quality and forest sustainability.

Acknowledgements

This work was undertaken as part of a Ph.D. project, jointly funded by the Macaulay Land Use Research Institute, Aberdeen, the Robert Gordon University, Aberdeen and the Scottish Office Agriculture, Environment and Fisheries Department. Thanks go to my supervisor, Dr. David Miller, for his help with the GIS data and ERDAS Imagine visualization software.

References

Berlyne, D.E. (1963) Complexity and incongruity variables as determinants of exploratory choice and evaluative ratings. *Canadian Journal of Psychology* 17: 274-290.

Bishop, I.D. and Hulse, D.W. (1994) Prediction of scenic beauty using mapped data and geographic information systems. *Landscape and Urban Planning* 30: 59-70.

Bishop, I.D. and Leahy, P.N.A. (1989) Assessing the visual impact of development proposals: the validity of computer simulations. *Landscape Journal* 8: 92-100.

Bishop, K. (1997) Picture perfect. *Demon Dispatches* 8: 45-48.

Brush, R.O. (1981) Landform and scenic preference: a research note. *Landscape Planning* 8: 301-306.

Daniel, T.C. and B. Orland (1997) Perceptual evaluation of SmartForest-II visualization system: comparison of analytic and landscape representations. *Data Visualization 97: Previewing the Future*. St Louis, Missouri, USA, October 1997.

Daniel, T.C. and J. Vining (1983) Methodological issues in the assessment of landscape quality. In: I. Altman and J. Wohwill (eds) *Behaviour and the Natural Environment*. Plenum Press, New York. pp. 39-83.

Dunn, M.C. (1976) Landscape with photographs: testing the preference approach to landscape evaluation. *Journal of Environmental Management* 4: 15-26.

ERDAS (1994) *Imagine, Field Guide*. ERDAS Inc., Atlanta, Georgia, USA.

Hammitt, W.E., M.E. Patterson and F.P. Noe (1994) Identifying and predicting visual preference of southern Appalachian forest recreation vistas. *Landscape and Urban Planning* 29: 171-183.

Hinkley, J. (1997) *Vistapro Version 4.01*. RomTech Inc., Langhorne, PA.

Hull, R.B. and G.R.B. Revell (1989) Issues in sampling landscapes for visual quality assessments. *Landscape and Urban Planning* 17: 323-330.

Hull, R.B., G.J. Buhyoff and T.C. Daniel (1984) Measurement of scenic beauty: the law of comparative judgment and scenic beauty estimation procedures. *Forest Science* 30: 1084-1096.

Imaging Systems Laboratory (1995) *SmartForest: an interactive forest data modeling and visualization tool*. Department of Landscape Architecture, University of Illinois at Urbana-Champaign.

Lange, E. (1998) *Reality and Computerized Visual Simulation: An Empirical Study About the Degree of Realism of Virtual Landscapes*. Institute of National, Regional and Local Planning, ETH Zurich.<http://www.orl.arch.ethz.ch/ ~Lange/schwyz/schwyz.html> 30/11/98 16:00.

Orland, B., E. Weidemann, L. Larsen and P. Radja (1995) *Exploring the Relationship between Visual Complexity and Perceived Beauty*. Imaging Systems Laboratory, Department of Landscape Architecture, University of Illinois at Urbana-Champaign. <http://imlab9.landarch.uiuc.edu/projects/complex/ >.

Pukkala, T. (1999) *MONSU forest modelling and simulation software*. Faculty of Forestry, University of Joensuu, Finland.

Shafer, E.L., J.F. Hamilton and E.A. Schmidt (1969) Natural landscape preferences: A predictive model. *Journal of Leisure Research* 1: 1-19.

Shuttleworth, S. (1980) The use of photographs as an environmental presentation medium in landscape studies. *Journal of Environmental Management* 11: 61-76.

Stamps, A.E. (1990) Use of photographs to simulate environments: a meta-analysis. *Perceptual and Motor Skills* 71: 907-913.

Wherrett, J.R. (1998) *Natural Landscape Scenic Preference: Techniques for Evaluation and Simulation*. Unpublished Ph.D. Thesis, Robert Gordon University, Aberdeen.

Wherrett, J.R. (2000) Creating landscape preference models using Internet survey techniques. *Landscape Research* 25: 79-96.

Wohlwill, J.F. (1968) Amount of stimulus exploration and preference as differential functions of stimulus complexity. *Perception and Psychophysics* 4: 307-312.

PART VI

**Reconciling Forest
Sustainability and
Aesthetics**

Chapter seventeen:

Priorities for Reconciling Sustainability and Aesthetics in Forest Landscape Management

S.R.J. Sheppard and H.W. Harshaw
Faculty of Forestry, University of British Columbia
Vancouver, British Columbia

J.R. McBride
Department of Landscape Architecture, University of California, Berkeley
Berkeley, California

So, what emerges from the musings of this eclectic group of scientists and authors? What can we say that is new and meaningful about the relationships between ecological sustainability and aesthetics in forested landscapes? Do people actually prefer sustainable landscapes - if not, can aesthetics and sustainability be reconciled, and how?

In the original Peter Wall Institute for Advanced Studies (PWIAS) Exploratory Workshop held at UBC in 1999, and in the introduction to this volume, we posed a number of questions around these issues. While there appears to be considerable consensus on some of these questions, across a range of disciplines among the Workshop participants and authors in this volume, the topic touches upon several far-reaching and largely unresolved areas of research. This leaves much room for argument and speculation.

In this final chapter, we try to map out the areas of consensus (Section 17.1), noting some of the substantive remaining differences in approach or opinion, and most importantly, we identify priorities for further research (Section 17.2). We hope that there are also lessons and implications here for practitioners seeking a more sensitive and perhaps, ultimately, more successful way to manage forested landscapes that are increasingly under public scrutiny. Our suggestions for forest managers are summarized in Section 17.3. The discussion in this chapter has been informed by the salient conclusions of this volume's individual authors, and from the insights that resulted from discussions at the PWIAS and the UBC Faculty of Forestry Exploratory Workshop, as described in Chapter 1.

17.1 Questions and Answers

It is inevitable that those of us who are interested in reconciling sustainable forest management with public perceptions of landscapes will be confronted with some broad questions. Some of these questions are reviewed first, as a background to the consideration of more specific issues of interest to forest managers, scientists, landscape architects, and other professionals and academics. In both cases, we attempt to provide concise answers reflecting the prevailing evidence and/or probabilities, although the opinions expressed here, including those characterizing areas of consensus, are those of the authors of this chapter; other authors in this volume may dissent on particular points.

17.1.1 Broad questions

(i) Do people prefer sustainable landscapes?

Yes, we believe that all things being equal, people do prefer sustainable forest landscapes. This is particularly true if the observers believe that the landscape they are seeing is being managed under sustainable principles. However, 'things' are often not equal.

Accepting some general definitions of what ecological sustainability means (*e.g.* those found in Chapters 4, 6, and 7), we can expect people to appreciate a healthy sustainable landscape if it matches certain biologically or culturally determined landscape preferences (Daniel, Chapter 2, this volume). We may also expect preference if the viewer is sure the landscape is sustainably managed, through certain knowledge or through apparent sustainability or visible stewardship (*i.e.* the landscape appears to be cared for in a sustained manner, even if the observer does not actually know whether the management is sustainable (Sheppard, Chapter 11, this volume)). However, where the sustainable landscape matches certain biologically or culturally determined preferences, or where people believe that a landscape's appearance is the result of an insensitive, non-collaborative process (see Kruger, Chapter 12, this volume) or from a primarily profit-driven venture, then people may not like what they see. However, when sustainable landscape conditions are met, and are neither conventionally scenic nor conventionally ugly, then it is unclear how important the role of information regarding the sustainability of management is in determining preferences.

(ii) Is there a relationship between aesthetics and ecological sustainability, and if so, what is the pattern?

Yes, we believe that a generally positive correspondence between aesthetics and ecological sustainability can be expected in theory in a forest landscape context, but only under the specific conditions suggested above.

Under these conditions, landscapes of higher or lower sustainability can be associated with higher or lower aesthetic quality. For example, it is easy to believe that if a landscape that has been subject to erosion is restored to a more healthy state, its aesthetic attractiveness will increase. However, the application of this relationship to other conditions will depend on many factors, including the time at which the aesthetic judgement was made, relative to the ecological stage of the managed forest (see Luymes, Chapter 13, this volume); the state of knowledge or belief of the observer; and other ecological forest landscape conditions.

No single overall relationship between aesthetics and ecological sustainability can be expected across the wide range of variation that may be encountered, even, for example, in a particular forest type such as the northern temperate forests. While some clear patterns of correspondence between the two sets of variables can be found, other patterns indicating inverse relationships are in evidence (see Plates 2, 3, and 17). Other examples of an inverse relationship include the much loved, but unsustainably irrigated, fertilized, weeded, and manicured lawns of parks in drier climates; and the unsightly retention of coarse woody debris for ecological purposes in harvested areas. In many other situations, we would not expect any theoretical relationship to apply, as many of the important ecological factors cannot be associated with visible results during the average person's life time (Kimmins, Chapter 4, this volume). Therefore, considerable uncertainty remains in attempts to map the complex relationships between visual and ecological parameters.

(iii) Are sustainable landscapes scenic?

Some sustainable landscapes are scenic, while other sustainable landscapes are not.

This question is really a narrower version of Question (i) above: it is confined to the examination of conventional scenic preferences for landscapes, but ignores other types of or variations in preference, as Daniel (Chapter 2, this volume) suggests. Under the conventional scenic aesthetic, we can identify sustainable landscapes that exhibit conventionally accepted scenic characteristics and other sustainable landscapes that do not. There has yet to be a compilation of comprehensive information on the relative frequency or abundance of these two extremes. It seems likely that the majority of forest landscapes will fall somewhere in between: somewhat sustainable, and somewhat scenic. Can we make more sense of this murky middle ground? Is it possible that, given enough time, all or most sustainable landscapes become scenic sooner or later, as in the case of natural fire disturbance?

(iv) Are scenic landscapes sustainable?

Some scenic landscapes are sustainable (see Plate 18), while other scenic landscapes are not.

This question is one of the corollaries to the preceding question. Again, some landscapes generally recognized as scenic are sustainably managed and seem to be relatively secure ecologically: North America's great wilderness parks serve as examples. However, there are other highly scenic landscapes that have proved to be unsustainable, such as the forests of the Lake Tahoe basin where previous logging, overly restrictive fire suppression, lack of disease control, and other factors have contributed to critical forest health issues. Again, the relative occurrence of these extremes on the sustainability spectrum, and of their intermediates has not been documented. We can speculate that unsustainable landscapes eventually lose their scenic value: this is what is happening in Lake Tahoe where the high percentage of standing dead trees has finally become noticeable to the public as an obviously unhealthy grey-brown backdrop to a shining blue lake.

This question also raises other issues about the management of scenic landscapes: What is the sustainability of various measures taken to protect aesthetic values? For example, can the use of helicopter logging (with the associated high rate of combustion of non-renewable fuels) to salvage trees or harvest in visually sensitive areas (in order to meet visual quality objectives) be considered sustainable? Taking this line of argument one step further, is it a sustainable practice to encourage the conventional mode for scenic appreciation, sightseeing from the private car? Promoting this activity forms the goal of major government funding initiatives such as the *National Scenic Byway Program* in the USA. Should we be encouraging more hiking, biking, and scenic transit ways in active forest management areas, as is the case in overcrowded and/or environmentally sensitive National Parks such as Yosemite and Denali? There appears to be little documented information on the trade-offs between these aspects of aesthetics and sustainability.

(v) Is it a problem if there is currently no general positive relationship between sustainability and aesthetics, or if the relationship is adverse, partial, weak or confused? Why should we care?

The authors, Workshop participants, and practitioners alike generally agree that the confused relationship between aesthetics and ecology is a problem, and one that we should care about. The problem arises because the public may like that which is not good for the environment, and may be getting the wrong impression of what is actually good resource management. There is also concern amongst some scientists that resource managers and forest scientists themselves may be wrong about what constitutes good management.

We suggest that there is no practical option other than further study of the topic; conflicts will continue to arise over precisely these issues, and managers need better science and guidelines to deal with them. Even if scientists fail to determine that there is a strong relationship between the appearance and the state of the landscape, the public appears to have done so; this perception is unlikely to change substantially without intense and sustained effort.

(vi) If we accept that it is undesirable for aesthetics and sustainability to be at cross purposes, how do we begin to reconcile aesthetics and sustainability? Can public education that focuses on explaining what is ecologically sustainable, and what is not, impart this knowledge, or are there other methods of achieving such a reconciliation?

There is clearly a strong feeling among many of the authors in this volume that a better public understanding of ecological conditions and processes is needed, if not to narrow the gap between aesthetic preference and ecological sustainability, then at least to improve public acceptance of ecological practices (Bell, 2000). This raises the issue of the difference between preference and acceptance, and the role of trade-offs in balancing aesthetic or ecological goals and values against other needs, such as water quality, safety, employment, and profit.

There would appear to be at least three levels at which a reconciliation of aesthetics and sustainability could take place, within the paradigm of improving public understanding.

1. Project level trade-offs. At the individual project level, such as a forest development plan, trade-off analysis would be aimed at the full disclosure of impacts and benefits to various resource values (including aesthetics and ecological values), and at integrated planning to optimize the combinations in the final decision. Examples of this approach include the ecological design that has been pursued in the Applegate watershed in Oregon (Apostol *et al.*, 2000) and current interdisciplinary landscape unit planning by UBC's Faculty of Forestry in the Arrow Forest District of BC under an Innovative Forest Practices Agreement (Arrow Forest License Group, 1999). There are many other precedents for enlightened collaborative planning which makes trade-offs between conflicting values, in various resource and land use planning studies (*e.g.* Friend and Hickling, 1997; British Columbia Integrated Resource Planning Committee, 1993; Brown *et al.*, 2000). Other examples can be found within the US Forest Service participatory processes for preparing National Forest Land Management Plans and Environmental Impact Statements under the National Environmental Policy Act (NEPA).

While such techniques can potentially provide a practical resolution to conflicts between aesthetics and ecological sustainability, it is not clear whether the participants in the process actually feel as though the conflicting values have been reconciled, or simply that the best accommodation has been struck. Carlson (Chapter 3, this volume) believes that reconciliation would occur within our aesthetic preferences: as he puts it: "our knowledge of the sustainability of a landscape can be more important to our aesthetic preferences than how it happens to look at any particular point in time." However, Daniel in Chapter 2 disagrees: "Ecological knowledge is probably not sufficient to produce an aesthetic preference for sustainable landscapes".

2. Broader extension, information dissemination, and public education. These
activities would be employed at local, regional, or national levels to explain
sustainable forest management and the degree to which it relates to what the public
sees and other societal values. New techniques such as interactive games involving
people in forest management scenarios (*e.g.* FORTOON (Kimmins and Scoullar,
1994)) promise to be particularly effective in improving awareness of ecological
needs; however, actual results in improving acceptance of sustainable management
strategies at a community or societal scale are hard to predict. Any such acceptance
at this wider level is less likely to be an actual change in aesthetic values, than it is
some degree of broad trade-off.

3. A fundamental cultural change in widely held aesthetic values. Led by Gobster
(see the Foreword to this volume), some have advocated a shift in the prevailing
cultural aesthetic, away from current perceptions of what is scenic and toward a
deeper awareness of ecological health and integrity. This is the level that poses the
trickiest questions for research and practice, which are pursued further below.

Historically, Western cultural perceptions of the values of landscapes have
tended first toward resource exploitation, and in recent decades toward style and
conventional aesthetics, rather than the creation of sustainable landscapes (Thayer,
1989). Gobster (1995) has suggested that the current *scenic aesthetic* is a result
of three cultural legacies: a preference for idealized nature; the predominance
of a static, visual mode of landscape appreciation; and an uneasiness associated
with change. Nassauer (1995) has also identified the predominance of this scenic
aesthetic in a study of urban landscapes, which determined that landscapes that
appear neat, orderly, and cared for are more likely to be deemed attractive by
the public. These legacies need to recognized, addressed, and modified so that
the development of a new *ecological aesthetic,* which seeks to encourage the
appreciation of sustainable landscapes that do not adhere to common notions of
beauty, can be fostered.

However, there is another possible paradigm for reconciling aesthetics and
ecological sustainability, which argues that it is not necessary to change people'
minds or adopt a whole new aesthetic. This alternative paradigm suggests that,
at least within a semi-natural, managed forest context (as distinct from more
intensively constructed and managed landscapes such as urban parks), forest
interventions or practices which are beautiful generally are good in the long term,
whereas what looks bad eventually becomes bad. For example, in British Columbia
during the 1970s, clearcut harvesting on steep slopes, the construction of extensive
road cuts, and side-casting later led to erosion and sedimentation; this indicates
that the worse things look, the worse the ecological effect can be. When applied to
less extreme situations, this argument suggests that it may be more a question of
time before the true ecological consequences show up, than a true conflict existing
between ecological and aesthetic values. If this were true, then the implications
of this are that scientists and ecologists may, at some level, have it wrong about
what is best for the long term forest environment; at the very least, scientists and

ecologists may not know enough to say definitively that they are right and that the public is wrong. To take another example, clearcutting and the messy practice of leaving considerable amounts of coarse woody debris in the harvest area may be bad, not because of the woody debris, but because of the scale or other attributes of the harvesting; in other words, the public may be right after all.

This alternative paradigm (*having faith in the public*) does not at first blush appear to stretch to situations such as fire suppression, where active elimination of fire, in part to protect aesthetic values, leads ultimately to a poorer environment. However, one could argue that over the long term the relationship holds (eventually it looks bad and is bad), or that perhaps the public perception should be trusted when the forest looks bad, but not necessarily when the forest looks good! The paradigm could also be modified, whereby the more extensive the departure of the intervention from natural processes/conditions, the uglier it ultimately gets (Plate 19). This paradigm would fit the more rapacious logging practices of the past, as well as the less visible fire suppression measures still practiced widely today. Inevitably, the definition of the time-scale being considered is key: even with cases of artificially devastated landscapes, such as the widespread denudation of large parts of the Lake Tahoe basin in the mining and logging eras of the 19th century (Goin, 1992). Nature's restorative powers can substantially heal both the ecosystem and the view.

Clearly, the current state of ecological and social science leaves ample room for such conjecture. One can even argue that, even when they agree, both the public and the forest scientists are wrong - for example, that in the very long term, large-scale beetle kill and fire are not harmful to the environment, only to our notion of good forest management.

Certainly in North America, we have very limited longitudinal information on which management practices work and which do not, over the long term of several forest rotations. While the preponderance of evidence over the shorter term would seem to be in the forest scientists' favour, the question of whether even the most objective and expert forest scientists really know what is best for the complex ecosystems, remains valid. Therefore it is important to acknowledge the considerable uncertainty lingering in the definition of sustainability.

(vii) Given the need for some humility among forest scientists, while accepting for a moment the feasibility of developing a new 'ecological' aesthetic, should we attempt to alter the dominant scenic aesthetic to take into account ecological factors? Furthermore, is it ethically justifiable to seek deliberately to sway public opinion?

The majority of the Workshop participants, across the disciplines represented, felt that the need for sustainability was paramount; therefore, the goal of changing society's prevailing aesthetic is justified. Nevertheless, in discussion of these issues both within and following the Workshop, academics from the social sciences and arts fields have raised serious questions about the ethics of seeking to convince the public to accept forestry practices which they feel to be bad. In part, this may arise from the substantial level of distrust which remains about the motivations of the forest industry and foresters in general.

Arguments in favour of a new aesthetic, raised by authors such as Botkin (Chapter 9, this volume), suggest that Western society's prevailing aesthetic has changed several times in recent centuries, reflecting changes in our understanding of the environment and in our interpretations of acceptability and beauty. Aesthetic appreciation, and the messages derived from landscapes, are likely to change even if scientists do not participate in the debate. The emerging reassertion of aboriginal world views on nature and landscapes provides a completely different type of aesthetic that is separate from the Western scenic and picturesque. This aboriginal aesthetic is wedded instead to the human ecology of survival and societal needs in a non-pastoral civilization, where the priority is to live in harmony with a landscape that was only partly under their control (see Umeek in Chapter 8); this perhaps provides an interesting precedent or model for the deeper ecological aesthetic suggested by Gobster.

Going beyond the themes of seeing and knowing that have been raised by Carlson (Chapter 3, this volume), it can also be argued that the general public is not single-minded, or simplistic in their appreciation of nature; and that people would respond favourably to additional information on sustainable futures for forested landscapes. Indeed, it is possible that withholding information drawn from the current state-of-the-art in forest ecology and management, for fear of changing people's minds, would in and of itself be unethical. Clearly, the ethical dimensions of this issue demand further attention, both in research and in practice.

(viii) If we accept that it is appropriate to encourage ecological appreciation in aesthetic preferences, is it actually feasible at the cultural level to foster an aesthetic that is rooted in ecology?

Gobster would suggest that such an aesthetic is feasible, and he is not alone in this belief. However, some of the authors of this volume contend that the development and acceptance of such an aesthetic would be very difficult, and cannot be achieved by education alone.

Following in the footsteps of Leopold (1968), Gobster (1995) believes that in order for the public to develop and appreciate an ecological aesthetic, they must experience the affected environment for themselves. Furthermore, sustainable landscapes must be explicitly interpreted to the public as such (Thayer, 1989; Gobster, 1995), in order to link experience and understanding in the appreciation of sustainable landscapes. The use of what Nassauer (1995) terms *cues to care*, or what Sheppard (Chapter 11, this volume) calls *visible stewardship*, may assist in the presentation and interpretation of sustainable landscapes; in addition to these principles, the development of a new or deeper aesthetic may require a variety of approaches, addressing all three of the levels described in question (vi) above (*i.e.* project decisions, generally improved awareness, and fundamental change in the cultural aesthetic). These approaches might include: strategic policy changes; endorsement of a comprehensive *visible stewardship* programme at a forest, state, provincial, or national scale; sustained extension activities; collaborative planning and trust-building at the local and regional scales; and supporting regulations and incentives. The ultimate effectiveness of such activities is not clear.

(ix) To what extent are prevailing aesthetic preferences culturally derived (and therefore perhaps more mutable), as opposed to being biologically pre-determined?

The successful development of a new aesthetic would depend in part on the answer to this profound question.

As Daniel (Chapter 2) and Sheppard (Chapter 11) demonstrate in this volume, the innate (biological) basis for aesthetic preference, explored in theories such as *Prospect-Refuge* (Appleton, 1996) and *Habitat Theory* (Orian, 1986), may set limits to the amount of change that is possible in aesthetic responses to naturalistic landscape conditions. Thus, a biological basis for aesthetic preference has the potential for major implications regarding the success of any strategy to modify prevailing aesthetic preferences. The fundamental question of the respective roles and weights of the biological versus cultural components of aesthetic preference as yet remains unanswered. Emerging research on the perceptions of First Nation and other aboriginal groups with differing perceptions of those landscapes valued by Western eyes, promises to yield fascinating results.

The following section considers some of the more detailed philosophical, research, and practical questions that arise from the general arguments outlined above and which have implications for scientists and managers.

17.1.2 Specific questions

(i) Can we define sustainable landscapes more quantitatively?

In order to establish an actual relationship between aesthetics and ecological sustainability, we need quantifiable indices for purposes of measurement and correlation with other key indices. There has been considerable research into defining what people like or find visually attractive, and time-honoured methods exist for measuring those preferences; there is much less agreement on quantifying sustainability.

Some of the problems in defining measures of sustainability, such as the dynamic nature of ecosystems and the multiple dimensions of sustainability, have been discussed in the Chapters by Kimmins (Chapter 4), Oliver *et al.* (Chapter 6), and Burley (Chapter 7). It is difficult to find agreement on what 'sustainability' is, let alone what specific measures should be used. Oliver *et al.* suggest some specific quantifiable types of indicators that can be aggregated together to address regional and national issues: these are helpful in the attempt to establish the much-needed robust indicators of sustainability, though a wider range of stand and landscape-specific indicators (beyond proportions of the forest in different structural stages) may need to be developed before we can expect to test for meaningful relationships with visual, aesthetic, or broader social factors. Burley's review of possible amenity-related indicators provides a further basis for development of more meaningful quantified measures (Chapter 7, this volume). The establishment of even just a few easily-distinguished levels of sustainability would help in analysing the relationship with aesthetic preference.

(ii) Why do most sets of sustainable forestry indicators ignore public perceptions and aesthetics?

In Chapter 7, Burley highlights the scarcity of hard indicators of aesthetic (and related social) qualities in forest management. This scarcity exists despite the relative abundance and professional acceptance of certain measures of aesthetic "success" in both perception research and practical visual resource management.

We can speculate that this situation stems from two main causes. First, many foresters have not traditionally received much education or practical training in social sciences, and in practice, many seem to have a very low comfort-level with human perceptions and aesthetic theory, or even on-the-ground aesthetic design practices. Social knowledge seems to percolate very slowly into the forestry profession: at UBC's Faculty of Forestry, for example, formal training in aesthetics will finally enter the required curriculum for the Bachelor of Science in Forestry degree in 2003, after more than 30 years of visual resource management being accepted in the management of public forests of North America.

Secondly, partly due to this gap in the education system, aesthetic issues and public perceptions are typically seen by foresters as something soft, fuzzy, intangible, subjective, non-scientific, and not appropriate for serious study. It is something that has been dealt with by a few experts, and often seen as a necessary evil to be borne rather than embraced by the forestry profession. It is therefore not surprising that these and other social indicators are not as fully developed in sustainability criteria indicators. Other professionals who are more comfortable with managing specifically for these values, such as landscape architects, planners, sociologists, and environmental psychologists, may not have been so effectively included in the process of setting standards and indicators of sustainability. It is also likely that some foresters feel that if the ecological conditions are well managed, the aesthetics naturally follows (as discussed further below), reducing the need for specific aesthetic indicators. Further, many foresters may feel justified in ignoring short-term aesthetic effects because, as discussed earlier, nature is likely to heal both the land and the view given sufficient time.

(iii) Is sustainability visible? Are there reliable, intrinsic, visible indicators of ecological sustainability?

Sheppard (1997) has explored the potential advantages of simple visual techniques for monitoring environmental conditions or management practices, using inexpensive ground-based photography. The science of applying remote sensing techniques to ground photography is developing (*e.g.* Clay, 1998). Some ecological indicators may be readily visible and measurable with photography, such as percent of dead trees or visual evidence of sedimentation; however, as Kimmins (in Chapter 4) emphasizes, many important indicators are not directly visible, at least not on a reasonable management time-scale when actions have to be taken before the dire consequences become visible. None the less, it is worth exploring the extent to which better ecological awareness, and specific training, could increase the number or accuracy of visible indicators of key forest health conditions; the trained forester sees much more than the casual forest visitor. There are precedents for developing enhanced public observation skills: for example, the campaign for identification and control of fruit flies in California, where brightly coloured fruit fly traps were placed in visible locations throughout settled areas and served as a reminder of the threat to the community; and the situation in the English Cotswolds in the 1970s where some local residents learned to identify early stages of Dutch elm disease before tree death occurred, at a time when sanitation exercises were still being recommended to halt the spread of disease.

(iv) Can we make forest stewardship more visible and if we do, will it help improve public acceptability of new forestry?

Are there stewardship activities which can enhance or go beyond the intrinsically visible aspects of sustainable forestry? In Chapter 11, Sheppard presents a range of possible techniques that could help to make visible more aspects of sustainable management. These obviously include designed *cues to care* (Nassauer, 1995) and explicit interpretation of management actions or inaction's, through mechanisms such as signs and other forms of publicity. However, Kruger (Chapter 12, this volume) argues convincingly that, in addition to landscape characteristics and the state of the viewer's knowledge about the state of the landscape's sustainability, the decision-making process for managing the landscape is key to people's preferences and judgements of acceptability. Additionally, true visible stewardship must deliver and make obvious a collaborative, open process for obtaining and applying community input into the forest management. If we concentrate only on the ecological aspects of sustainability, particularly if we do not effectively involve the public in the decision-making, then we are unlikely to achieve acceptance.

However, we do not yet know the feasibility of visible stewardship. Existing community forests in Ontario and British Columbia provide opportunities to begin measuring the cost and effectiveness of maintaining a visible local presence in ecologically and economically sustainable forest management (Allen and Frank, 1994; Duinker *et al.*, 1991).

(v) Do visual quality objectives help or hinder the achievement of ecosystem management goals such as biodiversity and ecological health, and vice versa?

The most common regulatory or design tool employed for aesthetic management in North America is the Visual Quality Objective (VQO), which determines how much visual change is allowed in forest management. VQOs are often viewed within the resource industries as scenic constraints on timber supply or other resource extraction potential (Picard and Sheppard, 2000), but how do they align with ecological objectives?

The evidence here is again contradictory. Researchers such as Bradley (1996) cite examples of mutually beneficial design measures which enhance aesthetics and wildlife habitat, for example; in current research in the Arrow Forest District of BC, we have found considerable overlap in both visually sensitive areas and domestic watersheds, which together call for delicate and limited harvesting practices. Government agencies such as the BC Ministry of Forests (BCMOF) assumes that some VQOs help meet ecological requirements in its calculations for the Timber Supply Review process (Picard and Sheppard, 2000). However, a common result of the traditional VQO approach, tied to more heavily used public viewpoints, is that industrial harvesting operations are pushed into the back-country where it is generally out of sight, but where impacts become concentrated on remaining old growth or other pristine ecosystems. Here then, the geographic scale over which ecological values and visual objectives are measured clearly affects the conclusions on any relationship.

(vi) Is visual resource management defunct? Is there a continuing role for it, given society's shifting priorities for more sustainable management?

While we may increasingly believe that scenic beauty is not the big issue, relative to ecological sustainability, we should be careful not to throw the baby out with the bath water. The visual appearance of the landscape transmits messages not only about its picturesque qualities and aesthetic enjoyment (narrowly defined), but also represents important meanings and symbolism for other social end environmental priorities. In the final analysis, it may represent the single best means of explaining to the public how sustainable is our forest management.

This explanation will increasingly require the careful application of design skills, taking visual and other considerations into account in an integrated (rather than a discipline by discipline) approach. Apostol *et al.* (2000) have applied aspects of integrated design. A thorough integrated design process will require the blending of the landscape architect's understanding of public perceptions with skilful design of feasable forest practices and a solid support of ecological science. Current numerically-based approaches to quantifying VQOs and visual impacts of forestry may not prove flexible or comprehensive enough to achieve good design solutions on their own. It is time to develop a new, more flexible and integrated paradigm for practical visual resource management, based on design to educate as well as to hide, and going beyond the laudatory but limited attempts of the US Forest Service to develop its Scenery Management System with ecosystem principles in mind (USDA Forest Service, 1997).

(vii) Will the adoption of an ecological aesthetic impose a scientific paradigm on a typically romantic tradition of the appreciation of landscapes?

Is ecological function destined to play the overbearing, ugly sister to the picturesque's Cinderella? The study of aesthetics comes from a philosophic tradition that has evolved from Baumharten's science of sensuous knowledge, to Hagel who has gave us our current conception of aesthetics: the appreciation of beauty as derived from both emotional reaction to the experience and our knowledge of the rules of appreciation. This aesthetic appreciation of forested landscapes may be closely related to what Wilson (1984) has termed biophilia, an affinity for nature, or the innately emotional affiliation of human beings to other living organisms. Wilson concludes that: "to the degree that we come to understand other organisms, we will place greater value on them, and on ourselves" (Wilson, 1984, p. 2). Umeek (Chapter 8, in this volume) describes the related belief in Tloo-qua-nah among the Nuu-chah-nulth people on Vancouver Island, as a spiritual binding together of the human and ecological communities.

However, the rich experience and deeper knowledge which ought to result from environmental experiences and spiritual relations with the forest may simply prove unattainable, or worse, not sufficiently attractive to many people with other priorities. There is the danger then, that in seeking to foster an ecological aesthetic, the awareness-building process becomes transformed into a more clinical process of education with cold, hard, objective scientific evidence. The effectiveness of education in influencing aesthetic preferences of forested landscapes has been demonstrated by Haider (1994). There is therefore, a risk that the Western romantic tradition characterized by imagination, emotion, and visionary experiences, or the spiritual worldview of traditional aboriginal communities, may be lost or weakened in people's everyday experience, if they begin consciously to analyse landscapes before judging them.

(viii) What is the role of visualization in these issues?

While much discussion of the potential of visualization in resource management focuses on the advantages and potential problems of its use in practice, there is also a potentially vital role to be played as a powerful research tool. The ability is now within our grasp to generate multiple visual simulations of different forest stand conditions at different times under different management routines, and at relatively high levels of realism, based on semi-automated procedures. This permits researchers to test systematically the effects of individual and composite/ factors on perceptions of forest management and the resulting landscapes. It seems likely that systematic visualizations which emulate reality convincingly, may become one of the standard media for research on the human dimensions of ecosystem management, and may in so doing help unlock the secrets of the basis of aesthetic perceptions: nature or nurture? Wherrett provides an example of this form of research in Chapter 16.

The discussions in the other chapters within Part V of this volume focus largely on the implications of emerging visualization tools in influencing decisions and public perceptions. Can these tools be used to inform people's perceptions of what a sustainable landscape looks like? Can human perceptions be regulated and shaped by visualizations?

Clearly, there is enormous scope here for improving people's awareness and understanding of whatever can be represented in visual form, such as perhaps the level of sustainability in a given landscape or the long term benefits or disadvantages of a current forestry practice. One can imagine visualizations being effective at all three levels of reconciliation between ecology and aesthetics: project-specific design and decision-making; the broader extension to the community through education and even "edutainment", and perhaps even in cultural change through fostering greater interest in sustainability from the typically less-involved general public. Virtual reality techniques seem to bring people through the door more effectively than any other tool with practical utility to the resource professional.

There is also, obviously, the potential for visualization procedures to improve the forest manager and scientist's own understanding of the forest, and their accumulating and increasingly complicated databases. The possibilities for enhancing collaborative planning, by providing more accessible and interactive pathways to learning and to tangible input to the design process for lay people, may promote more sensitive design of forest management activity, as described by Luymes in Chapter 13 of this volume, and by Sheppard (2000).

In their chapters, Luymes and Orland and Uusitalo have cautioned that as simulations become increasingly realistic and easier to automate, people will lose all ability to *see through* the simulation or suspend their belief in the imagery. The presentations of these realistic simulations (either as static images or dynamic 'movies') could be used as rhetorical devices, rather than as measuring tools or presentations of possible futures; this is where the danger lies, particularly where the rhetoric begins to diverge from a credible scientific basis.

The use of images to persuade or alter human perceptions is not a new phenomenon. In addition to many historical examples of propaganda, pop culture can offer some examples. In Baum's classic, *The Wizard of Oz* (1982), a 'common man' manipulates the perceptions of those around him by 'making believe' - masquerading as a wizard while hiding behind a screen. While the intentions of the wizard are arguably good, he finds himself in a position where he is making promises that he cannot keep: a heart for the tin man; a brain for the scarecrow; courage for the lion; and a return to Kansas for Dorothy. What gets the wizard into trouble is his trickery - making promises he can not keep. If we are able to demystify visualizations - step out from behind the screen - and encourage people to question the information provided, verify the visualization process and come to their own conclusions, then it may be possible to avoid the worst excesses of manipulation by imagery. Transparency of the medium is necessary. The intentions of the visualizations and simulations must not be covert, they must be made explicit to the viewer. As Orland and Uusitalo (Chapter 14, this volume) and Sheppard (in press) suggest, we need better principles and codes of ethics to govern the widening use of visualizations.

There are a number of particularly interesting features of emerging visualization methods which promise to generate new information relevant to forest sustainability and aesthetics, in both research and practice. Examples discussed by two of the world's leaders in landscape visualization, Brian Orland (Chapter 14) and John Danahy (Chapter 15) include:

- Immersive environments whereby the viewer's sense of being there is significantly enhanced through panoramic or head-mounted virtual reality images (see Figure 17.1). This promises to simulate (and predict) more realistically, the all-important emotional reactions of forest users.
- Animated travels through and over the forest can provide viewers with the freedom to roam and visualize from any location. In his talk at

Fig. 17.1. Panoramic visualization of forested landscapes in the Landscape Immersion Lab, Forest Sciences Centre, University of British Columbia, Vancouver.

the PWIAS workshop, Umeek (Dr. Atleo) spoke movingly of his first trip by helicopter over Clayoquot Sound as a co-chair of the Clayoquot Sound Scientific Panel: he described how he wept as the helicopter climbed above the trees and foreground ridges, and gradually revealed to him for the first time the true extent of the old growth logging in the backcountry which was his people's ancient domain. The ability of animated visualization to present such telling pictures, to see things from a different perspective literally, to look more deeply into or more comprehensively over the forest landscape - may provide more avenues for enlightening communities and forest managers alike.

- Demonstrations of the temporal dynamics of forest condition at the landscape scale will allowing for people to visualize future conditions and perhaps make better trade-offs between ugliness now and ecological health over the long-term, or vice versa.
- Using interactive graphic techniques or 'special effects' to 'peel away' the forest surface and disclose non-visible features that are important to sustainable management, *e.g.* soil productivity, early disease spread, hydrological processes, wildlife use areas, or key understory characteristics. Orland (1998) has described the SmartForest Program which provides some of these features for the use of forest planners in the US Forest Service.

17.2 Research Directions: An Agenda for the Marriage of Aesthetics and Ecological Function

Evidently, there has not been sufficient research to characterize overall patterns or relationships between aesthetics and ecological sustainability, or even map these within a subset of conditions and parameters. It is important to recognize that there is really a hierarchy of interlocking research questions which need to be addressed, with varying breadth of application: for example visual preference, versus overall aesthetic preference, versus general preferences. Similarly, how sustainability in the landscape is defined affects the questions to be explored.

However, as we have concluded above, these types of question continue to be worth asking, and in fact need to be answered. There was consensus among PWIAS Workshop participants that there is a need for more cross-disciplinary research linking the study of aesthetics and public preference with the study of forest sustainability. While some may believe that there is no relationship between ecology and aesthetics at all (*i.e.* one would expect examples of coincidence and mutual exclusivity specifically because there is no overall relationship), there seems to be enough evidence that there are sets of conditions where there is a direct relationship (levels of ecological condition can be correlated with aspects of aesthetic quality). We primarily need to be able to say when and where these relationships can be expected to occur, and where they do not apply.

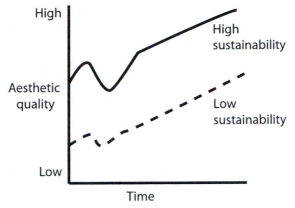

Fig. 17.2. Postulated relationships between ecological sustainability, aesthetic quality and time.

Ultimately, the questions posed in this volume need to be reframed for research purposes. If we stay within the focus of aesthetic preferences and ecological sustainability, one of the key questions is how the relationships between these factors develop over time. Time may be the key to unlocking any true relationships. As Luymes points out in Figure 13.1, there is a drastic dip in aesthetic quality and enjoyment when conventional timber harvesting takes place in a semi-natural forested environment. Applying this model to the issue of sustainability, we can imagine a graph relating aesthetic quality to forest sustainability over time, as shown in Figure 17.2 (Sheppard, 1999). For any given ecological condition, aesthetic quality may vary over time, especially due to harvesting activities; however, we can hypothesize that the average aesthetic quality over time may be considerably lower with less sustainable landscapes, than it would be with more sustainable landscapes.

Key research questions in this context therefore include:

- Determining the size, duration, and nature of the *aesthetic dip* over a meaningful number of sustainable forestry rotations, across different management scenarios (*e.g.* more and less sustainable harvesting techniques) within a particular landscape type;
- Determining the average or prevailing level of aesthetic quality across management scenarios (*e.g.* more and less sustainable management practices) for a given landscape type;
- Determining if there is a relationship between the nature of the *dip* and the level of sustainability;
- Determining if there is a relationship between the average or prevailing level of aesthetic quality and the level of sustainability; and
- Determining methods for shortening and minimizing the *aesthetic dip* in more sustainable forms of forest management.

A vital component of any research design along these directions is to account for the effects of scale, from individual stands to more complicated temporal and spatial patterns of harvesting and management at the landscape scale. It is also becoming increasingly important to gain a better understanding (and perhaps rules of thumb for practitioners) of the lengths of time into the future, over which people demonstrate concern for forest conditions and aesthetics. If the *aesthetic dip* is too long-term relative to people's typical residency periods in a community or relative to their lifetime, then do their concerns for longer-term sustainability still remain intact?

While the PWIAS Workshop participants agreed that sole reliance on visible indicators of forest health is limiting, it is also clear that we need to explore more and better ways to making true forest sustainability issues tangible. Testing the effectiveness of visualization methods in explaining these more subtle realities has been discussed at length above. Testing the validity of these tools in comparison with real world landscapes is also an urgent priority, to underpin the emerging paradigm of visualization-based perceptual research. Better monitoring and evaluation of the use of visualizations in forestry communications and decision-making is sorely needed.

In addition, the following suggestions for research priorities to link aesthetics and sustainability have been put forward:

- *The need for an ecological landscape classification system*, similar to what Klinka and Krajina developed for BC's biogeoclimatic zones (Krajina, 1976) but which more explicitly links physiographic, vegetation, and lands use features to visual characteristics of the landscape.
- *Application of statistical tools, such as principle component analysis, to assist in the identification of appropriate criteria and indicators of ecological change*; which explicitly seeks to identify where relationships to the visual landscape occur.
- *Refinement and testing of stand growth and succession models* that incorporate natural disturbance and human activities (such as the Landscape Management System (McCarter *et al.*, 1995) and FORECAST, FORECEE, and HORIZON (Kimmins *et al.*, 1999)), where physiographic characteristics of landscape are tied in to predict landscape change over time. Once established, these links can provide the basis for the development of landscape-sensitive models by landscape architects and visualization experts, with development of temporal visualization tools driven by ecological models rather than by fixed and partially supported estimates or snapshots.

Beyond research to map out the relationships between aesthetics and ecological sustainability in practical and experimental conditions, there is need for the study of the fundamental mechanisms behind human perception of sustainable forest management, including investigating the roots of aesthetics: biology or culture?

Creative research designs need to be developed to tease apart the relative weight of these factors, in cross-cultural studies of landscape perception.

A related topic that is worthy of study is the aesthetics of the forest scientist and forest manager. Is fragmentation of the forest really bad, or does it just look bad to scientists raised on theories of landscape ecology and island biogeography? Studies of perceptions of forest fragmentation by foresters (*e.g.* D'Eon and Glenn, 2000) have begun to emerge. Indeed there are fascinating side-issues: if there is general agreement on the level of fragmentation based on rapid visual inspection of aerial views, do we in fact already have a robust indicator of the actual ecological impacts of fragmentation? It would be instructive to apply more studies to foresters, like those carried out by Nassauer (1988) on farmers, to see if there are any aesthetic tendencies analogous to the farmers' preference for order and neatness as a sign of good husbandry.

In summary, we can define the primary research directions in these areas as follows:

- Research to map the relationships between aesthetics and sustainability.
- Research to explore the fundamental causes and theories of aesthetics as it relates to sustainability.
- Research to develop more robust indicators of sustainability and site classification, including more explicit links with values which are socially and aesthetically meaningful.
- Research to examine the nature of scientific and professional perceptions of sustainability, and how these differ from stakeholders, lay people's and other cultural groups' views.
- Research to develop effective means of displaying and communicating the dynamic nature of ecological processes.
- Research to validate and evaluate forest landscape visualization methods.
- Research to establish variables that could describe landscape visualizations systematically, so that visualizations that have been developed using different techniques can be compared.
- Research to automate the creation of geometric-based data, especially the texturing of extensive geographic surfaces with trees.

17.3 Some Implications for Forest Management

In this section, the authors suggest certain implications of the preceding discussions for the practice of forest management. Given the patchy and uncertain state of knowledge which prevails on the relationships between sustainability and aesthetics, we would not presume to attempt comprehensive listing of recommendations. However, certain truths and strong indications have emerged from the debate so far, and are worthy of consideration by practitioners, pending further corroborating research. Indeed, current forestry practice may provide the most conclusive evidence of all on these difficult questions.

For the typical forest manager seeking to reconcile public acceptance with his or her other objectives (*i.e.* obtaining what is often called a social licence), it would make sense to focus first on those temporal stages and more extreme conditions where the contrasts between the visible scene and sustainable practice are most dramatic: *i.e.* harvesting time (the short-term *aesthetic dip*), and perhaps certain natural or semi-natural disturbances such as fire or advanced infestation. However, this is not to say that reconciling aesthetics and sustainable practices at a forest-specific or project-specific level can be achieved in reactive mode at the time that these critical periods occur. If trade-offs are to be made between looking good and doing good, if minds are to be changed on the acceptability of certain ecologically beneficial practices, or if ecological goals are to be over-ridden by aesthetic needs, considerable advanced planning is required. Open minds and trusting relationships are usually not created overnight.

Based on the more robust answers to the questions raised in this volume, and drawing upon both research evidence and practical experience, the following summary of recommendations for practice is offered:

- Do not assume that aesthetic considerations will be met by providing for the other resource values, especially ecological values. As Jacques Marc, Head of the BCMoF Visual Landscape Management Program, told the PWIAS, practices such as the retention of woody debris, even-width stream buffers, wildlife patches, and shelterwood harvesting often lead to public reactions as negative as those generated by clearcutting, unless landscape design considerations are carefully built in. Thus, advance planning for timber supply or specific harvesting operations needs to consider such design issues explicitly at an early stage, and much greater use of the landscape architect or visual specialist's skills should be made in an integrated planning/design process. This is necessary to build into the design process both creativity and a working knowledge of the near-universal patterns of public response to certain visual interventions in the forest, which have been well-documented in perceptual research (referenced in several places in this volume), but are largely unknown to most foresters. One could argue that all community forests should employ a landscape architect to conduct visual management and community-based design, and generally to help to achieve the best fit for revenue-generating forestry with the local landscape.
- Do not assume that even with 'good' ecological and landscape design, there is no need for a substantive and open public input process. As Bell (2000) suggests planners and resource managers should concentrate more on the cognitive factors and social processes which affect public reactions to forest management practices, rather than just on the design of practices themselves. This is supported by Carlson's arguments for the importance of the viewers' knowledge in opinion forming on the landscape (Chapter

3), and Kruger's call for the importance of inclusive of social processes in securing public acceptance (Chapter 12). As Tindall points out in Chapter 5, "one cannot simply measure things in the woods to understand human and social values associated with forests". This perhaps helps to explain for example, why the introduction of the radically different Forest Practices Code designed to safeguard biodiversity and environmental values in BC has failed to stem the flow of protest against forest harvesting. It may also explain why certification of a forest management programme based largely on forest practices and company performance may miss the boat in achieving a locally-derived social licence. At the same time, there is no point in providing a public input process if the decision-makers are not prepared to listen to the results, capitalize on local knowledge, and adapt their plans accordingly.

- Develop an active programme of *visible stewardship*, where the image and reality are synchronous, and the sustainable practice of forestry is made plain within the community. Sheppard in Chapter 11 suggests means of building this image through highly visible actions, a continuous presence within the affected community, a clear attachment to and respect for the land as *place* rather than commodity, building of environmental awareness, and a visible public involvement process which seeks to apply collaborative community design methods. Practitioners could, for example, experiment with radical landscape designs which express the uniqueness of the local environment or community in some new way, while demonstrating a commitment to sustainable practices.

- Use defensible landscape visualizations not only as a final presentation tool, but also as an interactive planning and design tool during the process. These need not be limited only to issues of visual impact: they can be used and adapted to present much important information which other media cannot as effectively portray. They can also be used to provide the vital temporal pictures of future conditions, with which the community can judge the appropriate trade-offs. Practitioners should also, however, insist on ethical and verifiable visualizations, to avoid real or apparent bias in their use of simulations.

As discussed above, there are implications for the use of VRM in forest management. VRM practitioners have unique and important skills which need to be applied and disseminated throughout the forestry profession (Strain and Sheppard, 1997). Visual design deserves to be more fully integrated with ecological design and other resource considerations, not by 'feel' but by balancing scientifically-derived quantitative outcomes and community-derived preferences in the light of improved ecological information. The old models of specialized, separate, pseudo-quantitative VRM programmes need to be replaced with more flexible but ultimately more powerful and acceptable systems which aim to do more than simply hide forestry from the public. It is the visual specialist's particular challenge to find the means to an effective policy of visible stewardship, and to use the landscape itself as a major information source promoting sustainability. The unique perceptions of aboriginal forest users, such as the First Nations in Canada, integrating spiritual and other cultural values into their view of the forest, need to be fully engaged in a fuller forest planning process, drawing on emerging research results on these topics. Here too, the skills and training of the landscape professional can be uniquely valuable in helping to bridge the gap between conventional Western science and traditional knowledge and customs.

The visual alteration of forested landscapes is not a purely aesthetic consideration; increasingly there is an economic consideration as well: resource dependent communities seek to diversify their economies through the pursuit of *ecotourism* and the promotion of recreation activities. The interplay of these two resource dependent activities will require more attention by forest managers as tourism grows, back-country use expands, and associated international awareness continues to increase. Good forest stewardship may increasingly need to be made visible at a considerable distance, using techniques such as web access to real-time monitoring imagery and virtual tours of landscape units or harvesting operations.

Finally, there is a need for:(i) more integration of other disciplines in forest planning strategies, *e.g.* sociologists, landscape architects, alternative resource economists, *etc.*, along with the more traditionally involved ecologists; and (ii) a better and more rounded education of foresters in the area of human perceptions and aesthetic principles. Forestry programmes need to provide more exposure to social contexts generally, and reduce the sometimes yawning gap between the public and the forester.

17.4 Summary Discussion and Conclusions

While it may be right to expect some correspondence between aesthetics and ecology, it can be seen that what is believed to be ecologically good may not look good (Plate 3), and that what looks good may not be ecologically sustainable (Plate 17) (Nassauer, 1995; Sheppard, Chapter 9, this volume).

The authors of this volume for the most part share a consensus that it is vital to reconcile conflicts between aesthetics and sustainability in order to foster more sustainable management of forests and forest-dependent communities; and that it is ethical to provide the public with better information on ecological outcomes, and perhaps even contribute to a new public aesthetic which accepts the concept of forest dynamics and enlightened human intervention. Cultural aesthetic norms have changed, sometimes radically, over the centuries and even in shorter periods. It is appropriate that they should continue to evolve as new information becomes available and society's priorities shift. However, we should be watchful that these newer perceptions reflect not only the new information from traditional science in fields such as forestry, but also the emerging worldviews and local knowledge sources of other forest cultures and the growing diversity of forest users.

The quest for an *ecological aesthetic* seeks to reconcile human impressions of beauty with functioning, sustainable ecosystems. Too often in the recent past, human endeavour has pursued the picturesque or the profitable at the cost of sustainable forested landscapes. As many of the authors in this volume have argued, ecosystem function and beauty need not be mutually exclusive, indeed they have not always been viewed as such. However, practices which in some ways emulate large-scale natural disturbance, *e.g.* clearcutting, have been cause for public outcry and has fuelled the dissension over timber extraction on an industrial scale, in part due to the dramatic imagery and associated visual messages that result.

The current *scenic aesthetic* prevalent in the appreciation and understanding of forested landscapes is at least in part tied to cultural legacies that do not address ecological sustainability. These legacies include a preference for idealized nature, the predominance of a static and visual mode of landscape appreciation, and uneasiness associated with change; this aesthetic is built into, and perhaps even perpetuated by, current forms of Visual Resource Management of forests. This scenic aesthetic does not adequately address the dynamic nature of landscapes. We urgently need a deeper understanding of temporal change in forests, and the associated public understanding and perceptions. New techniques of computer-based landscape visualization will play a pivotal role in allowing us to 'turn the dial forward' to possible forest futures, and measure for the first time people's expected responses to forest conditions over time, as well as the trade-offs they are prepared to make (or not) for the sake of long term sustainability.

The development and acceptance of a new *ecological aesthetic* is one way to ensure that ecosystems in the public eye are managed sustainably. It has been suggested that in order for this ecological aesthetic to be nurtured, there must be a change in the way people appreciate landscapes, based on experiencing and understanding the ecological processes necessary for the ecological sustainability of natural landscapes. However, it is considered likely by the authors that much more will be needed to effect a comprehensive reconciliation of aesthetics and sustainability for forestry. Major shifts in approaches to forest planning, such as improved decision-support approaches for socially-inclusive trade-off analysis, and making forest landscape stewardship more visible, will be required.

The authors of this volume have recognized the need for new approaches such as the development of an ecological aesthetic and visible stewardship, and it is hoped that this book helps to lay the groundwork necessary for such a development. This concern and appreciation for sustainable landscapes has resonated with the public as demonstrated by the interest in the PWIAS Workshop from practitioners and the public alike. Clearly, much more research and testing in practice is needed to resolve the questions raised in this book, including the development of more robust hypotheses and empirical evidence describing the patterns and relationships between forest conditions and human responses.

Pragmatically, if the development of new cultural aesthetics via a public education approach is uncertain, there may be other ways of expressing more clearly the sustainability of resource management. Ultimately, the importance of the visual landscape as a symbol of ecological health and quality of forest management mandates that we find ways to make the sustainable more visible.

Acknowledgements

The public speakers at the PWIAS Workshop who have contributed to this volume include: Jacques Marc, Senior Resource Specialist in the Forest Development Section at BC Ministry of Forests; Bill Cafferata, formerly the Vice President and Chief Forester for MacMillan Bloedel, currently a member of the British Columbia Forest Practices Board; Tim Howell, attorney, Sierra Legal Defense Fund, Vancouver; and Dean Apostol, consultant and landscape architect, Oregon. We would also like to recognize the substantive contribution to the outcome of the Workshop and hence to these conclusions, of the other Workshop participants, namely Wolfgang Haider, Assistant Professor at the School of Resource and Environmental Management, Simon Fraser University; Walt Klenner, Wildlife Biologist, Research Unit, BC Ministry of Forests, Kamloops, BC; and Dr. Kellogg Booth, Director of Media and Graphics Interdisciplinary Centre, Computer Science, UBC.

References

Allan, K. and D. Frank (1994) Community forests in British Columbia: Models that work. *Forestry Chronicle* 70(6): 721-724.

Apostol, D., M. Sinclair, and B. Johnson (2000) Design your own watershed: Top-down meets bottom up in the Applegate Vallley. In: R.G. D'Eon, J. Johnson, and E.A. Ferguson (eds), *Ecosystem Management of Forested Landscapes: Directions and Implementation.* Ecosystem Management of Forested Landscapes Organizing Committee, Vancouver, BC, pp. 45-50.

Appleton, J. (1996) *The Experience of Landscape.* John Wiley and Sons, Chichester.

Arrow Forest License Group (1999) *Innovative Forestry Practices Agreement: Forestry Plan.* Unpublished report.

Baum, L.F. (1982) *The Wizard of Oz.* Holt, Rinehart and Winston, New York.

Bell, S. (2000) Ensuring the social acceptability of ecosystem management. In: R.G. D'Eon, J. Johnson, and E.A. Ferguson (eds), *Ecosystem Management of Forested Landscapes: Directions and Implementation.* Ecosystem Management of Forested Landscapes Organizing Committee, Vancouver, BC, pp. 69-84.

Bradley, G.A. (1996) *Forest Aesthetics - Harvest Practices in Visibly Sensitive Areas.* Washington Forest Protection Association, Olympia, WA.

British Columbia Integrated Resource Planning Committee (1993) *Land and Resource Management Planning: A Statement of Principles and Process.* Victoria: Integrated Resource Planning Committee, 21 pp.

Brown, K., W.N. Adger, E.B.P. Tompkins, D. Shim, and K. Young (2000) *Trade-*

off Analysis for Marine Protected Area Management. GEC Working Paper 2000-02. Centre for Social and Economic Research on the Global Environment, University of East Anglia, Norwich, UK, pp. 32.

Clay, G. R. (1998) Integrating spatial data with photography for visualizing changes in a forested environment. *Journal of the Urban and Regional Information Systems Association,* Fall.

D'Eon, R.G. and S.M. Glenn (2000) Perceptions of landscape patterns: Do the numbers count? *Forest Chronicle* 76(3): 475-480.

Duinker, P.N., P.W. Matakala, and D. Zang (1991) Community forestry and its implications for Northern Ontario. *Forestry Chronicle* 67(2): 131-135.

Friend, J. and A. Hickling (1997)*Planning Under Pressure:The Strategic Choice Approach*, 2nd ed. Butterworth Heinemann, Oxford.

Gobster, P.H. (1995) Aldo Leopold's ecological esthetic: Integrating esthetic and biodiversity values. *Journal of Forestry* 93(2): 6-10.

Goin, P. (1992) *Stopping Time: A Rephotographic Survey of Lake Tahoe*. University of New Mexico Press.

Haider, W. (1994) The aesthetics of white pine and red pine forests. *Forestry Chronicle* 70(4): 402-410.

Kimmins, J.P. and K.A. Scoullar (1994) Incorporation of nutrient cycling in the design of sustainable, stand-level, forest management systems using the ecosystem management model FORECAST and its output format FORTOON. In: L.O. Nilsson (ed) *Nutrient Uptake and Cycling in Forest Ecosystems.* Ecosystem Research Report Number EUR 15465.

Kimmins, J.P. (Hamish), B. Seely, D. Mailly, K.M. Tsze, K.A. Scoullar, D.W. Andison, and R. Bradley (1999) FORCEEing and FORECASTing the HORIZON: Hybrid simulation modeling of forest ecosystem sustainability. In: A. Amaro and M. Tome (eds) *Empirical and Process-Based Models for Forest Tree and Stand Growth Simulation*. Edicoes Salamandra, Lisboa, Portugal, pp. 431-441.

Krajina, V.J. (1976) *Biogeoclimatic Zones of British Columbia*. University of British Columbia Faculty of Forestry, Vancouver, British Columbia.

Leopold, A. (1968) *A Sand County Almanac and Sketches Here and There*. Oxford University Press, London.

McCarter, J. B., J. Wilson, M. Wimberly, J. Moffett and C.D. Oliver. (1995). A Landscape Management System. *IUFRO XX World Congress – Caring for the Forest: Research in a Changing World, 6-12 August*. Tampere, Finland.

Nassauer, J.I. (1988) The aesthetics of horticulture: neatness as a form of care. *Horticultural Science* 23: 937-977.

Nassauer, J. (1995) Messy ecosystems, orderly frames. *Landscape Journal* 14: 161-170.

Orians, G.H. (1986) An ecological and evolutionary approach to landscape aethetics. Chapter 2 In: E.C. Penning-Rowsell and D. Lowenthal (eds) *Landscape Meanings and Values*. Allen and Unwin, London.

Orland, B. (1998) *Smart Forest-II: Forest Visual Modeling for Forest Pest*

Management and Planning. Report under USDA Forest Service Cooperative Agreement 28-C2-602. Urbana, Illinois: University of Illinois, Department of Landscape Architecture, Imaging Systems Laboratory. Http://imlab.uiuc.edu/ SF/Sfcompon.html.

Picard, P. and S.R.J. Sheppard (2000) Timber supply and aesthetics at the landscape level: Theoretical relationships and potential win-win solutions. *Journal of Ecosystems and Management* (Southern Interior Forest Extension and Research Partnership). In press.

Sheppard, S.R.J. (1997) Photo-monitoring: a tool for measuring landscape change. *In: IAIA '97 Proceedings. International Association of Impact Assessment*, New Orleans, March 1997

Sheppard, S.R.J. (1999) *The Visual Characteristics of Forested Landscapes: A Literature Review and Synthesis of Current Information on the Visual Effects of Managed and Natural Disturbances.* Report prepared for BC Ministry of Forests Research Division, Kamloops and Adams Lake IFPA. 40pp.

Sheppard, S.R.J. (2000) Visualization as a decision-support tool in managing forest ecosystems. *Compiler* 16(1): 25-40.

Sheppard, S.R.J. Guidance for crystal ball gazers: Developing a code of ethics for landscape visualization. *Landscape and Urban Planning* (special issue). 14p. (In press).

Strain, A. and S.R.J. Sheppard (eds.) (1997) *Workshop Proceedings, Establishing the Worth of Scenic Values: The Tahoe Workshop, Lake Tahoe, October, 1997.* <http://www.ceres.ca.gov/trpa/SCENIC.PDF>.

Thayer, R.L. (1989) The experience of sustainable landscapes. *Landscape Journal* 8(2): 101-110.

United States Department of Agriculture Forest Service (1997) *Landscape Aesthetics: A Handbook for Scenery Management.* Agriculture Handbook #701. USDA Forest Service, Washington, DC.

Wilson, E.O. (1984) *Biophilia.* Harvard University Press, Cambridge, MA.

Index

Armstrong
SD
387
S87
F69
2001